PROBLEM SOLVING/
THINKING MA

Algebra 1

HOLT, RINEHART AND WINSTON

A Harcourt Classroom Education Company

Austin • New York • Orlando • Atlanta • San Francisco • Boston • Dallas • Toronto • London

To the Teacher

Problem Solving/Critical Thinking Masters contain blackline masters that provide a variety of problem-solving strategies for each lesson of *Algebra 1*.

Each master includes a description of a problem-solving strategy and how it applies to a specific situation. Students receive instruction in each problem-solving strategy by viewing a completed example. Students then practice the problem-solving strategy in a set of exercises. At times, a master can provide needed practice in previously introduced problem-solving strategies. Most masters include critical-thinking questions that require students to draw conclusions, make hypotheses, and assess their problem-solving skills.

Photo Credit
Front Cover: (background), Index Stock Photography Inc./Ron Russell; (bottom), Jean Miele MCMXCII/The Stock Market.

Printed in the United States of America

ISBN 0-03-054289-8

7 8 9 10 066 07 06 05 04

Table of Contents

Problem Solving/Critical Thinking

1.1 *Using a Table to Continue Number Sequences*

One method of analyzing sequences is to subtract pairs of adjacent terms until you find a constant difference. A table can be used to organize the work.

◆ **Example**

Find the next three terms of this sequence by using constant differences.

317, 233, 166, 114, 75, 47, 28, . . .

◆ **Solution**

Since you need to find three more terms, add three blank boxes to the top row of the table. Complete the white part first from top to bottom. Then complete the shaded part from bottom to top. Each box shows the difference of the numbers in the two boxes above it.

317		233		166		114		75		47		28		16		9		5
	84		67		52		39		28		19		12		7		4	
		17		15		13		11		9		7		5		3		
			2		2		2		2		2		2		2			

Complete the tables to show the first five terms of each number sequence.

1.
183		100		47		18		
	83							

2.
139		88		52		28		
	51		36					

Find the next three terms of each sequence. Draw tables to help you organize your work.

3. 240, 210, 182, 156, 132, 110, 90, . . .

_____

4. 246, 203, 164, 129, 98, 71, 48, . . .

_____

5. 979, 710, 495, 328, 203, . . .

_____

6. 858, 610, 414, 264, 154, . . .

_____

7. **Critical Thinking** At the right is an arrangement of numbers called *Pascal's Triangle*. Describe how Pascal's Triangle is related to the method of constant differences which you have been using on this page.

```
                    1
                  1   1
                1   2   1
              1   3   3   1
            1   4   6   4   1
          1   5  10  10   5   1
        1   6  15  20  15   6   1
      1   7  21  35  35  21   7   1
```

_____

_____

Problem Solving/Critical Thinking

1.2 Using Guess-and-Check to Solve Equations

One way to solve equations is to use the guess-and-check strategy. In this strategy, you can try any number that you choose for the variable. Then you adjust your guess to come closer to the solution until you finally find it. Here are some tips for using guess-and-check.

◆ **Tips**

Remember that the expressions on both sides of the equation must be equal. Use this fact to decide whether your first guess should be positive or negative. For your first guess, choose a number that will make the computation easy. Analyze the equation to decide whether your next guess should be larger or smaller.

◆ **Examples**

$\frac{2n}{3} - 6 = 2$

Use a multiple of 3, such as 3, 6, or 9, so that you won't need to subtract fractions.

Try 9: $\frac{2 \cdot 9}{3} - 6 = \frac{18}{3} - 6 = 0$

Since $0 < 2$, the value of $\frac{2n}{3} - 6$ is too small.

Try a larger number, such as 12 or 15.

$23 - 3x = -19$

Use a positive number so that the left side of the equation will be negative.

Try 10: $23 - 3(10) = 23 - 30 = -7$

$-7 > -19$, so the left side of the equation must be smaller. Therefore, the value of $3x$ should be larger. Try a larger number, such as 12 or 14.

For each equation, explain how you would choose your first guess.

1. $54 + 6x = 10$ _____

2. $\frac{4}{5}x + 22 = 78$ _____

3. $4 + \frac{3}{x} = 8$ _____

4. $\frac{6 \cdot 2}{1 - x} = 3$ _____

For each equation, a first guess is given. Determine whether the guess is too small or too large.

5. $4x - 37 = -5$
First guess: 5

_____

6. $57 + 3x = 6$
First guess: -10

_____

7. $\frac{x + 3}{-4} = 2$
First guess: -7

_____

8. $17 - \frac{x}{6} = 8$
First guess: 36

_____

9. Critical Thinking Explain what information you can gain from an incorrect guess to a problem.

_____

_____

NAME _____ CLASS _____ DATE _____

Problem Solving/Critical Thinking

1.3 *Evaluating Expressions Step-by-Step*

Evaluating expressions requires that you use the algebraic order of operations. An organized list can be used to transform the work into a step-by-step sequence. You may find that this method works better for you than rewriting the expression at each step.

♦ **Example**   Evaluate this expression.

$$\dfrac{3 \cdot 6^3 - 48}{4(57) + 6(14 - 2)}$$

♦ **Solution**   Make an organized list to plan and show your computation.

In the numerator, compute 6^3.	$6^3 = 216$
Multiply by 3.	$216 \cdot 3 = 648$
Subtract 48.	$648 - 48 = 600$
In the denominator, subtract 2 from 14.	$14 - 2 = 12$
Multiply by 6.	$6 \cdot 12 = 72$
Multiply 4 times 57.	$4 \cdot 57 = 228$
Add the two products.	$228 + 72 = 300$
Write the final fraction.	$\dfrac{600}{300}$
Simplify this fraction.	$600 \div 300 = 2$

Write the first two steps that you would use to evaluate each expression.

1. $15 + 8^2 - 3 \cdot 24 + 7 - (4)(38)$

2. $(6 - 3)(29) + (8 + 5)(30 - 21)$

3. $(8)(26) - (32)(27 - 6) + 52$

4. $48 + 3 \cdot 8^4 - (73 + 4 \cdot 2)$

5. $\dfrac{28 + 9 \cdot 6}{(82 - 7^2) \cdot 16}$

6. $\dfrac{(3)(7 - 2) + (8 - 4)(3)}{(11 + 3)(14 - 1)}$

Write a complete plan for evaluating each expression.

7. $(16 - 9)(23) - 3^5 \cdot (6 + 4 \cdot 8)$ _____

_____

8. $\dfrac{5 + 4^3 - 11 \cdot 2}{(51 + 6)(3 \cdot 2 + 4)}$ _____

_____

9.   **Critical Thinking**   Use the expression $\dfrac{9 \cdot 4 - 28 - 2^3}{30 \cdot 48 - 23^2}$ to answer the following:

a.   Evaluate the numerator step-by-step. Describe the steps you use. _____

b.   Why is it not necessary to evaluate the expression in the denominator? _____

_____

Problem Solving/Critical Thinking

1.4 Analyzing Graphical Models

When using a graph to solve a problem, start by looking at the graph to get a rough idea of how the two quantities are related.

◆ **Examples**

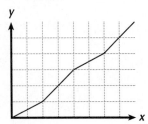

As the value of x increases, the value of y also increases.

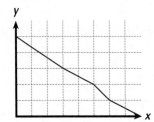

As the value of x increases, the value of y decreases.

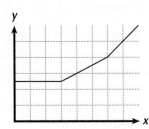

As the value of x increases, the value of y stays the same and then increases.

Use each graph for the problems next to it.

1.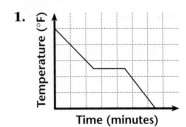

    a. What happens as the number of minutes increases? _____

    _____

    b. When was the temperature at a maximum? _____

    _____

2.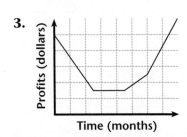

    a. What happens as the number of pounds increases? _____

    _____

    b. What happens as the cost decreases? _____

    _____

3.

    a. What happens as the months pass? _____

    _____

    b. When did this store start to make a profit?

    _____

4. **Critical Thinking** Describe a graph that shows no apparent relationship between the two variables.

_____

_____

Problem Solving/Critical Thinking

1.5 *Using First Differences*

When you are asked to write an equation for a data set, you may find it helpful to imagine a situation that can be described by the data.

For the data set shown, the first differences are all 4. The data could represent an item that costs $4 each, such as a book. When $x = 0$, $y = 7$. Seven dollars could represent an admission fee.

0	1	2	3	4	5	6
7	11	15	19	23	27	31

Therefore, one example that could model the data is a museum that charges $7 admission and then sells books for $4 each. This situation can be represented by the equation $y = 4x + 7$.

Find the first differences, and describe a situation that might be modeled by the given data. Then write an equation to represent the data pattern.

1.

0	1	2	3	4	5	6
60	70	80	90	100	110	120

First differences: _____

Equation: _____

Situation: _____

2.

0	1	2	3	4	5	6
12	17	22	27	32	37	42

First differences: _____

Equation: _____

Situation: _____

3.

0	1	2	3	4	5	6
0	8	16	24	32	40	48

First differences: _____

Equation: _____

Situation: _____

4.

0	1	2	3	4	5	6
0	3	6	9	12	15	18

First differences: _____

Equation: _____

Situation: _____

5.

0	1	2	3	4	5	6
8	9	10	11	12	13	14

First differences: _____

Equation: _____

Situation: _____

6. **Critical Thinking** For Exercise 5, write another possible situation that does not involve money.

_____

Problem Solving/Critical Thinking
1.6 Using a Table to Analyze Data

You can make a scatter plot to see if there is a correlation between two sets of
data. Such a correlation, if it exists, can also be determined from a table.

◆ **Example**
   Determine whether there is a correlation between these two sets of test scores.

Student	A	B	C	D	E	F	G	H	I	J
Math scores	75	96	59	53	99	83	97	77	83	83
Science scores	62	95	60	45	96	85	94	75	74	74

◆ **Solution**
   Reorganize the data, putting the math scores in increasing order.

Student	D	C	J	A	H	I	F	B	G	E
Math scores	53	59	83	75	77	83	83	96	97	99
Science scores	45	60	74	62	75	74	85	95	94	96

   The science scores, in general, are increasing, so there is a positive
   correlation between the two sets of data.

**Reorganize the data in each table to determine whether there
appears to be a correlation. If there is a correlation, state
whether it is positive or negative.**

1.
Data set A	11	5	7	5	8	8	6	3	13	9
Data set B	16	24	20	25	22	18	25	28	14	19

   Is there a correlation? _____

2.
Data set A	37	34	56	45	37	39	53	42	48	35
Data set B	22	58	67	12	4	41	11	75	43	9

   Is there a correlation? _____

3.
Data set A	426	378	358	394	345	437	402	362	418
Data set B	144	178	186	163	209	140	152	174	158

   Is there a correlation? _____

4. **Critical Thinking** Use the test score data at the top of this page.
   What conclusion would you reach if you put the science scores rather
   than the math scores in increasing order?

   _____

   _____

Problem Solving/Critical Thinking

2.1 Using a Number Line to Sort Real Numbers

When you compare two real numbers, you express a relationship between them using <, >, or =. When a list contains more than two real numbers, you may find a number line to be a useful visual aid in sorting them from least to greatest.

◆ **Example:** Sort the list below from least to greatest.
 2.5, 1.5, 4, −2, −5, −0.5

◆ **Solution:** Draw a number line such as the one below.

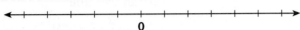

Group the negative numbers to the left of 0 and the positive numbers to the right of 0.
 −2, −5, and −0.5   2.5, 1.5, and 4

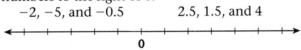

Choose a negative number, such as −3. Compare each negative number to it. Choose a positive number, such as 3. Compare each positive number to it.
 −5 < −3   −2 > −3   −0.5 > −3   2.5 < 3   1.5 < 3   4 > 3

Compare the negative numbers greater than −3.
Compare the positive numbers less than 3.
 −2 < −0.5   1.5 < 2.5

Finally, gather all this information to make the sorted list.
 −5, −2, −0.5, 1.5, 2.5, 4

On the number line given, group the negative numbers and the positive numbers in each list.

1. 3, −3, 0.5, −2.5, 4

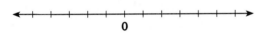

2. 4, 3.5, 1.5, −1.5, −2.5

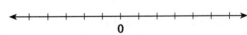

3. $\frac{1}{2}$, 3, $-\frac{5}{2}$, 4, $-1\frac{1}{2}$

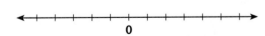

4. 1.2, −2.2, 3, −0.5, $1\frac{1}{4}$

5. Using your number line from Exercise 3, continue sorting

and write the list in order from least to greatest. _____

6. Critical Thinking Let $x > 0$. Sort the list shown at right so that the expressions will be in order from least to greatest. $-|x - 2|, -|x + 3|, |x + 4|, -|x + 4|, |x + 5|$

_____

Problem Solving/Critical Thinking

2.2 *Number Line Models for Integer Addition*

You have seen how to use algebra tiles to add integers. Here is the number-line method. This method works well for numbers with relatively small absolute values.

◆ **Example**
Find the sum of $-7 + 5$ on a number line.

◆ **Example**
Find the sum of $2 + (-8)$ on a number line.

◆ **Solution**
To add a positive integer, move to the right on the number line.

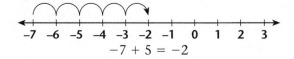

$$-7 + 5 = -2$$

◆ **Solution**
To add a negative integer, move to the left on the number line.

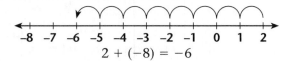

$$2 + (-8) = -6$$

Write the expression and the sum for each number line model.

1.
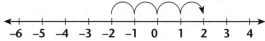

_____

2.
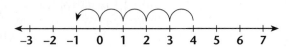

_____

3.
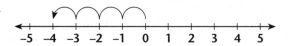

_____

4.
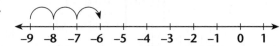

_____

Use a number line model to find each sum.

5. $-3 + (-2)$ _____

6. $4 + (-6)$ _____

7. $8 + (-5)$ _____

8. $-5 + 8$ _____

9. **Critical Thinking** Explain how you could use a number line numbered by 10s for a problem such as $-53 + 87$. Use the space below to make a number line that models your explanation.

_____

Problem Solving/Critical Thinking

2.3 Using Diagrams for Integer Subtraction

Many times you can use a skill that you have mastered in order to deal with a problem that is new. Because subtraction is defined in terms of addition, you can transfer what you know about addition to the concept of subtraction.

◆ **Example**
Draw a diagram to solve $2 - 5$.

◆ **Solution**
Change $2 - 5$ to $2 + (-5)$, an addition problem. Then use a number line.

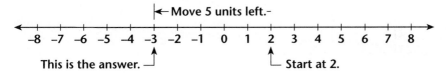

Rewrite each of the following as addition problems. Then use a diagram to find the solution.

1. $-2 - 3$

   = _____ + _____

   = _____

2. $4 - (-2)$

   = _____ + _____

   = _____

3. $-3 - (-5)$

   = _____ + _____

   = _____

4. $-6 - (-6)$

   = _____ + _____

   = _____

5. $-8 - 2$

   = _____ + _____

   = _____

6. **Critical Thinking** Describe how to use a vertical number line to add or subtract integers.

_____

_____

Problem Solving/Critical Thinking

2.4 *Patterns in Integer Multiplication*

The patterns in a multiplication table can help you understand the rules for multiplying integers. In the whole number table at right, notice that each row increases in a regular pattern from left to right.

×	1	2	3	4
1	1	2	3	4
2	2	4	6	8
3	3	6	9	12
4	4	8	12	16

1. Complete this multiplication table.

×	−4	−3	−2	−1	0	1	2	3	4
−4									
−3									
−2									
−1									
0									
1									
2									
3									
4									

2. Shade or color the table to show that it has four different sections or quadrants.

3. What parts of the table show each of the following?

a. the product of two negative integers _____

b. the product of two positive integers _____

c. the product of two integers with the same sign _____

d. the product of two integers with opposite signs _____

4. Explain how the table illustrates multiplication of integers.

_____

5. Use the table to predict the value of (−6)(3). Explain your thinking.

_____

6. Critical Thinking Describe how the table could be used to explain division of integers.

_____

Problem Solving/Critical Thinking

2.5 *Solving a Simpler Problem: Looking for 0 and 1*

Many times a problem may look more complicated than it really is. If you make observations when you look at a problem, you may find that the problem is simpler than you thought. When you examine an expression, notice any pairs that have 0 as their sum or 1 as their product. You may find that you can solve a simpler problem.

◆ **Example**
$$-3 + 6 - (-3) - 9$$
$$= -3 + 6 + 3 - 9$$
$$= \boxed{-3 + 3} + 6 - 9 \longleftarrow \text{sum of 0}$$
$$= 6 - 9 \qquad \longleftarrow \text{simpler problem}$$

◆ **Example**
$$0.25 \cdot 3 \cdot 0.5 \cdot 4$$
$$= \frac{1}{4} \cdot 3 \cdot 0.5 \cdot 4 \quad \longleftarrow \text{Recognize } 0.25 = \frac{1}{4}.$$
$$= \boxed{\frac{1}{4} \cdot 4} \cdot 3 \cdot 0.5 \longleftarrow \text{product of 1}$$
$$= 3 \cdot 0.5 \qquad \longleftarrow \text{simpler problem}$$

Rewrite each expression and circle sums of 0. You do not need to finish the computation.

1. $4 - (-2) - (3 + 1) + 2$ **2.** $3 - (7 - 2) + 3 + 5 - (-2)$ **3.** $6 + (-4) - (-6 + 2) - 2$

_____   _____   _____

4. $-5 + (-2) - (-4 + 2) + 3 - (-4) - 2$ **5.** $-(-6 + 10) - 4 + 14 - (6 - 2) + 4$

_____   _____

Rewrite each expression and circle products of 1. You do not need to finish the computation.

6. $0.4 \cdot 5 \cdot 4 \cdot 0.2$ **7.** $0.5 \cdot 0.25 \cdot 5 \cdot 2$ **8.** $8 \cdot 0.5 \cdot 4 \cdot 0.125$

_____   _____   _____

9. $0.8 \cdot 0.4 \cdot 1.5 \cdot 0.6 \cdot 1.25$ **10.** $2 \cdot \frac{2}{3} \cdot 3 \cdot 1.5 \cdot 0.5$ **11.** $\frac{1}{7} \cdot \frac{7}{8} \cdot 1\frac{1}{8} \cdot 1\frac{1}{7} \cdot \frac{1}{8}$

_____   _____   _____

12. $2 \cdot 0.25 \cdot 5 \cdot 0.1 \cdot 0.6 \cdot 4 \cdot 0.2 \cdot 3$ **13.** $\frac{4}{5} \cdot \frac{1}{4} \cdot \frac{2}{5} \cdot 1\frac{1}{2} \cdot 1\frac{1}{5} \cdot 1\frac{1}{4} \cdot 2\frac{1}{2} \cdot \frac{1}{2} \cdot \frac{3}{5}$

_____   _____

14. **Critical Thinking** Adding 0 does not change the value of an expression. How is this idea used in the following equation? Do you think that this simplifies the computation?

$$62 - 79 = 62 + (-2 + 2) - 79 + (-1 + 1)$$

_____

_____

Problem Solving/Critical Thinking

2.6 *Simplifying Expressions with Tables*

When you are first learning to add and subtract expressions, you may find it helpful to organize like terms in columns. First, write the expression as a sum.

◆ **Example**

$(3n + 5) - (6y - 2n) + (-4 - 2y) = 3n + 5 + (-6y) + 2n + (-4) + (-2y)$

	n-terms	*y*-terms	constant terms
	3*n*	−6*y*	5
	2*n*	−2*y*	−4
sum	5*n*	−8*y*	1

$\rightarrow 5n + (-8y) + 1 = 5n - 8y + 1$

First, write each expression as a sum. Then simplify by using the chart.

1. $(-3x - 4) - (2 - x)$

	x-terms	constants
sum		

= _____

= _____

2. $(-1 + b) + (-2b + 4) - (1 - 3b)$

	b-terms	constants
sum		

= _____

= _____

3. $(p + 3) + (4 - 2q) - (4q - 2p)$

	p-terms	*q*-terms	constants
sum			

= _____

= _____

4. $-(c - 2d) - (d + 1) + (3c - 2d) - (3 + 4c)$

	c-terms	*d*-terms	constants
sum			

= _____

= _____

5. Critical Thinking In the space at the right, create a different kind of table for simplifying expressions. Use the type of term for the entries in the left column rather than for the top row. Show how you would work Exercise 4 with this new kind of table.

_____

_____

Problem Solving/Critical Thinking

2.7 *Solving a Simpler Problem: Avoiding Fractions*

Part of algebra is learning to write expressions in different forms. You can use this to your advantage to simplify computation. Many students do not like working with fractions. If you are one of them, you can rewrite many expressions to avoid fractions.

◆ **Example**

$$\frac{8x^2 - 12}{-4} = (8x^2 - 12) \div -4 = (8x^2 \div -4) + (-12 \div -4)$$
$$= -2x^2 + 3$$

Simplify each expression by rewriting it as a quotient rather than a fraction.

1. $\dfrac{2x + 8}{-2}$

2. $\dfrac{-12w^2 + 4}{2}$

3. $\dfrac{9 - 3y^2}{-3}$

4. $\dfrac{15a - 20}{-5}$

5. $\dfrac{-6 - 9p}{3}$

6. $\dfrac{-d^2 - (-3)}{-1}$

7. $\dfrac{3x^2 - 6x + 3}{-3}$

8. $\dfrac{-4y^2 + 8y - 20}{2}$

9. $\dfrac{2b^2 - 3b + 4 - b - 10b^2}{4}$

10. $\dfrac{-17 + n^2 + 3n - 3 - (11n^2 - 2n)}{-5}$

11. **Critical Thinking** Dividing by a number is the same as multiplying by its reciprocal. Use this idea to simplify the following expression. Show your steps at the right.

$$\frac{3x^2 - 6x + 9}{\frac{1}{3x}}$$

Problem Solving/Critical Thinking

3.1 Predicting Solutions to Equations

You can probably predict certain things about the solution before you begin to solve. Then you can use your predictions to check to see whether your answer makes sense.

◆ Examine the equation $x + 4 = 7$. Notice that 4 is added to an unknown number, x, to get 7, a larger number. Therefore, x must be a positive number less than 7.

For each equation, check all the boxes that describe the solution. Do not solve the equations.

1. $a - 7 = 9$ The value of a will probably be
- ☐ a fraction.
- ☐ a whole number.
- ☐ greater than 9.
- ☐ less than 9.

2. $n + 2.3 = 8$ The value of n will probably be
- ☐ a whole number.
- ☐ a decimal.
- ☐ greater than 8.
- ☐ less than 8.

3. $\frac{2}{3} + x = 4$ The value of x will probably be
- ☐ a fraction less than 1.
- ☐ a fraction greater than 1.
- ☐ a positive number.
- ☐ a negative number.

4. $18 - d = 25$ The value of d will probably be
- ☐ an integer.
- ☐ a fraction.
- ☐ a positive number.
- ☐ a negative number.

5. $0.5 = b + 6.2$ The value of b will probably be
- ☐ an integer.
- ☐ a decimal.
- ☐ a positive number.
- ☐ a negative number.

6. $4 = y - 12$ The value of y will probably be
- ☐ a positive number.
- ☐ a negative number.
- ☐ greater than 12.
- ☐ less than 12.

7. $\frac{7}{9} = \frac{1}{3} - f$ The value of f will probably be
- ☐ a positive fraction.
- ☐ a negative fraction.
- ☐ less than $\frac{-1}{3}$.
- ☐ less than $\frac{-7}{9}$.

8. $-7 = q + 4$ The value of q will probably be
- ☐ a positive integer.
- ☐ a negative integer.
- ☐ greater than –7.
- ☐ less than –7.

9. Critical Thinking Write at least two predictions about the solution of each equation.

a. $14 - n = 1.5 - 8$ _____

b. $\frac{5}{6} + \frac{3}{4} = h - \frac{1}{2}$ _____

Problem Solving/Critical Thinking

3.2 *Predicting Solutions to Equations*

You can often predict characteristics of a solution before you solve an equation. When the equation involves multiplication or division, apply these key ideas.

Sign of the answer	Size of the answer
If you multiply or divide two numbers with the same sign, the answer is positive.	If you multiply by a number greater than 1, the product will be greater than the first factor. If you multiply by a number less than 1, the product will be less than the first factor.
If you multiply or divide two numbers with different signs, the answer is negative.	If you divide by a number greater than 1, the quotient will be less than the dividend. If you divide by a number less than 1, the quotient will be greater than the dividend.

For each equation, check all boxes that describe the solution. Do not solve the equations.

1. $\frac{a}{3} = 8$ The value of a will probably be
   - ☐ a whole number.
   - ☐ a fraction.
   - ☐ less than 8.
   - ☐ greater than 8.

2. $13t = 2$ The value of t will probably be
   - ☐ a whole number.
   - ☐ a fraction.
   - ☐ less than 13.
   - ☐ greater than 13.

3. $\frac{m}{5} = 1.3$ The value of m will probably be
   - ☐ a whole number.
   - ☐ a decimal.
   - ☐ greater than 5.
   - ☐ less than 5.

4. $-3w = 2.4$ The value of w will probably be
   - ☐ a positive number.
   - ☐ a negative number.
   - ☐ less than –2.4.
   - ☐ greater than –2.4.

5. $0.2 = 6x$ The value of x will probably be
   - ☐ a whole number.
   - ☐ a decimal.
   - ☐ less than 0.2.
   - ☐ greater than 0.2.

6. $\frac{4}{5} = \frac{m}{-3}$ The value of m will probably be
   - ☐ a positive number.
   - ☐ a negative number.
   - ☐ less than $-\frac{4}{5}$.
   - ☐ greater than $-\frac{4}{5}$.

7. **Critical Thinking** Write at least two predictions about the solution of the equation $16 - 2 \cdot 4 = -5n$.

   _____

   _____

Problem Solving/Critical Thinking

3.3 *Alternate Solution Methods*

There is often more than one way to solve a problem. Working a problem two different ways can help you check your answer and choose the best method for a future problem.

When solving a two-step equation, you must "undo" both operations in the equation. Here are two ways to solve the equation $2x + 1 = 12$.

◆ **Method 1**

Undo addition and then undo multiplication. Subtract 1 from each side to get
$$2x = 11.$$
Divide both sides by 2 to get:
$$x = \frac{11}{2} = 5\frac{1}{2}$$

◆ **Method 2**

Undo multiplication and then undo addition. Divide each side by 2 to get
$$x + \frac{1}{2} = 6.$$
Subtract $\frac{1}{2}$ from both sides to get:
$$x = 5\frac{1}{2}$$

Solve each equation in the two indicated ways. Show your work.

1. **a.** $\frac{x}{5} + 7 = -2$

   Subtract and then multiply.

   b. $\frac{x}{5} + 7 = -2$

   Multiply and then subtract.

2. **a.** $3x - 9 = 27$
   Add and then divide.

   b. $3x - 9 = 27$
   Divide and then add.

3. **a.** $\frac{x}{2} - 4 = 3$
   Add and then multiply.

   b. $\frac{x}{2} - 4 = 3$
   Multiply and then add.

4. **a.** $10 + 5x = 20$
   Subtract and then divide.

   b. $10 + 5x = 20$
   Divide and then subtract.

5. **Critical Thinking** Explain how to decide which step to do first when solving a two-step equation.

_____

_____

_____

Problem Solving/Critical Thinking

3.4 *More Alternate Solution Methods*

As you begin to solve more complicated equations, you will find that there are many ways to solve each equation. As long as each of your steps leads towards simplifying the equation or isolating the variable on one side, you are probably on the right track. However, some methods may be easier than others.

◆ **Example**   Here are three ways to solve the equation $\frac{x}{4} - 2 = \frac{x}{5} + 4$.

$$\frac{x}{4} - 2 = \frac{x}{5} + 4$$

$$20\left(\frac{x}{4} - 2\right) = 20\left(\frac{x}{5} + 4\right)$$
$$5x - 40 = 4x + 80$$
$$5x - 40 - 4x = 4x + 80 - 4x$$
$$x - 40 = 80$$
$$x - 40 + 40 = 80 + 40$$
$$x = 120$$

$$\frac{x}{4} - 2 = \frac{x}{5} + 4$$

$$20\left(\frac{x}{4} - 2\right) = 20\left(\frac{x}{5} + 4\right)$$
$$5x - 40 = 4x + 80$$
$$5x - 40 + 40 = 4x + 80 + 40$$
$$5x = 4x + 120$$
$$5x - 4x = 4x + 120 - 4x$$
$$x = 120$$

$$\frac{x}{4} - 2 = \frac{x}{5} + 4$$

$$\frac{x}{4} - 2 + 2 = \frac{x}{5} + 4 + 2$$

$$\frac{x}{4} - \frac{x}{5} = \frac{x}{5} + 6 - \frac{x}{5}$$

$$x\left(\frac{1}{4} - \frac{1}{5}\right) = 6$$

$$x\left(\frac{1}{20}\right) = 6$$

$$x = 120$$

Solve each equation at least two different ways. Show your work.

1. $3x + 4 = 8x - 16$

2. $5a + 10 = 2a + 25$

3. $-4n + 7 = 5n - 11$

4. $6s - 5 = 11s + 25$

5. $\frac{x}{3} - 5 = \frac{x}{4} + 2$

6. $\frac{z}{5} - 4 = \frac{z}{3} + 6$

7. Critical Thinking Explain how to decide which step you should do first when you solve a multistep equation.

_____

_____

Problem Solving/Critical Thinking

3.5 *Recognizing Special Solutions*

Most of the time, a linear equation will have just one solution. But, as you learned in Example 3 of this lesson in your textbook, some linear equations have all real numbers as solutions and other equations have no real solutions.

You can recognize these special cases by simplifying both sides of the equation.

- If both sides of the simplified equation are the same, all real numbers are solutions.

- If the x-terms are the same on each side but the constant terms are different, there are no real solutions.

- If the x-terms are different or an x-term appears on only one side, there is just one solution.

Simplify both sides of each equation. Then describe the solution to the equation by writing *one real solution, no real solution,* or *all real numbers.*

1. $3(2x + 5) = 2(3x - 4)$

_____

_____

2. $5p - 3 + 2p = 18 + 4p$

_____

_____

3. $2k - 6 - 8k = -6(k + 1)$

_____

_____

4. $5(3x + 4) - 6x = 3(3x + 2)$

_____

_____

5. $3y - 5(y + 2) = -2(y + 5)$

_____

_____

6. $-2m + 4(m + 3) = -2(m - 6)$

_____

_____

7. $3.6(z - 6) = 7.2 - 3.6z$

_____

_____

8. $2.4(5c - 4) = 7c - 5(1 - c)$

_____

_____

9. Critical Thinking Explain why there are no real solutions if the x-terms are the same on each side but the constant terms are different.

_____

_____

Problem Solving/Critical Thinking

3.6 Checking Solutions to Literal Equations

You know how to check your solutions when solving equations in one variable. You can perform the same kind of check when solving literal equations. Just substitute your answer for the appropriate variable in the original equation, and make sure that a true equation results.

Example Solve for l and check: $P = 2l + 2w$

Solution

$$P = 2l + 2w$$
$$P - 2w = 2l + 2w - 2w$$
$$P - 2w = 2l$$
$$\frac{P - 2w}{2} = \frac{2l}{2}$$
$$\frac{P - 2w}{2} = l$$
$$l = \frac{P - 2w}{2}$$

Check

$$P \stackrel{?}{=} 2\left(\frac{P - 2w}{2}\right) + 2$$
$$P \stackrel{?}{=} (P - 2w) + 2w$$
$$P \stackrel{?}{=} P + (-2w + 2w)$$
$$P = P \qquad \text{True}$$

Solve each equation for the indicated variable, and check your solution.

1. $k + l = m$ for k

2. $d - h = j$ for h

3. $3x + 2y = z$ for x

4. $5x - 4y = 3z$ for x

5. $f = ma$ for m

6. $I = prt$ for t

7. $w = fg + h$ for g

8. $d - ef = g$ for f

9. $y = 5x + 8z$ for x

10. $y = 4z - 3x$ for x

11. $5p - 4q = r$ for q

12. $6r + 2s = 7$ for r

13. Critical Thinking If you produce a false equation when checking a solution, what do you think has happened, and what should you do?

_____

_____

Problem Solving/Critical Thinking

4.1 Labeling Quantities in Proportions

When you set up a proportion to solve a problem, it is important that both ratios compare quantities in the same order.

◆ **Example** Susan earned $35 in 5 hours. At this salary, how much will she earn in 12 hours?

Here are 16 *different* possible proportions for this problem.

$$\frac{35}{5} = \frac{n}{12} \quad \frac{12}{5} = \frac{n}{35} \quad \frac{35}{n} = \frac{5}{12} \quad \frac{12}{n} = \frac{5}{35} \quad \frac{5}{35} = \frac{12}{n} \quad \frac{5}{12} = \frac{35}{n} \quad \frac{n}{35} = \frac{12}{5} \quad \frac{n}{12} = \frac{35}{5}$$

$$\frac{35}{5} = \frac{12}{n} \quad \frac{12}{5} = \frac{35}{n} \quad \frac{35}{n} = \frac{12}{5} \quad \frac{12}{n} = \frac{35}{5} \quad \frac{5}{35} = \frac{n}{12} \quad \frac{5}{12} = \frac{n}{35} \quad \frac{n}{35} = \frac{5}{12} \quad \frac{n}{12} = \frac{5}{35}$$

The proportions in the first row will result in the correct answer, but the proportions in the second row will not. With so many possibilities, you can see that it would be very easy to write a wrong proportion for the problem.

◆ **Solution** Write words or units in the proportion in order to ensure that you are comparing the quantities in the same order.

$$\frac{\text{dollars}}{\text{number of hours worked}} \longrightarrow \frac{35 \text{ dollars}}{5 \text{ hours}} = \frac{n \text{ dollars}}{12 \text{ hours}}$$

Set up a proportion using words or labels. Do not solve the proportion.

1. Art ran 13.4 miles in 2.25 hours. At this speed, how long will it take him to finish a marathon that is 26.2 miles long? _____

2. A 200-foot building casts a shadow that is 3 yards long. How tall is another building that has a shadow that is $5\frac{1}{2}$ yards long? _____

3. In a reference book, Matt found that 1 liter equals about 0.22702 gallons. How many liters are there in a 40-gallon tub? _____

4. It took Sharon 4.2 hours to drive 85 miles. At this rate, how long will it take her to complete a 145-mile trip? _____

5. Trisha read 78 pages of a history book in 2.25 hours. How many pages can she read in 5 hours? _____

6. A family spent $525 for food in January. If this represents an average amount, how much will they spend during the entire year? _____

7. **Critical Thinking** Four people can do a job in 8 days. The supervisor learns that the job must be done in 2 days rather than 8 days. So, he puts 16 people to work on the job. Did the supervisor use a proportion to solve this problem? Explain your answer.

_____

_____

Problem Solving/Critical Thinking

4.2 Estimating with Percents

Estimating the answer to a percent problem can help you write the
equation or proportion that you need to solve the problem. Starting with
the original problem, write related statements using easy percents, such as
50%, 25%, and 10%, and easy numbers, such as 100 and 200.

◆ **Example**
What number is 43% of $789?

◆ **Example**
$85 is 78% of what number?

◆ **Example**
$7.85 is what percent of $65?

◆ **Solution**
25% of $800 is $200.
50% of $800 is $400.
The answer should be
somewhat greater than $300.

◆ **Solution**
$85 is 50% of $170.
$85 is 100% of $85.
The answer should be
between $85 and $170.

◆ **Solution**
$6.50 is 10% of $65.
$13 is 20% of $65.
The answer should be
between 10% and 20%.

**Write at least one related statement for each problem. Then
estimate the solution.**

1. What number is 23% of 69?

   _____

   _____

2. 12.4 is 11% of what number?

   _____

   _____

3. 132 is 148% of what number?

   _____

   _____

4. 13.7 is what percent of 64?

   _____

   _____

5. 61 is what percent of 12?

   _____

   _____

6. What number is 5.8% of 146?

   _____

   _____

7. What number is 123% of 72?

   _____

   _____

8. 82 is 98% of what number?

   _____

   _____

9. 16 is 4.2% of what number?

   _____

   _____

10. 0.28 is what percent of 0.67?

    _____

    _____

11. **Critical Thinking** Ellen starts every percent problem by thinking,
    "Ten percent of 100 is 10." Why is this not a good problem-solving
    strategy?

    _____

Problem Solving/Critical Thinking

4.3 *Tree Diagrams for Listing Outcomes*

In solving probability problems, you may need to find the number of distinct outcomes. One way to do this is to draw a tree diagram. Often you do not need to draw the entire diagram to find the number of outcomes.

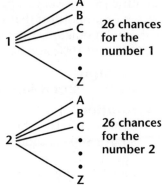

◆ **Example**

In a school raffle, the numbers 1 through 100 and the letters A through Z are placed in two boxes. One number and one letter are drawn at random. You win a prize if your age and either your first or last initials are drawn. What is the probability that Sam Gianaro will win a prize?

◆ **Solution**

Start the tree diagram as shown. It illustrates that the number of possible outcomes is $100 \cdot 26$, or 2600. Because Sam's first and last initials are different, he has 2 chances out of 2600 of winning. The probability is $\frac{2}{2600} = \frac{1}{1300}$, or about 0.08%

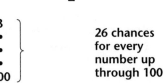

Find the number of distinct outcomes for each situation. In any situation where items are drawn, the first item is replaced before the second is drawn.

1. Two 6-sided number cubes are rolled.

   _____ outcomes

2. A penny and a quarter are tossed.

   _____ outcomes

3. A spinner with 4 equal parts labeled with the letters P, L, A, and Y. The spinner is spun

   twice. _____ outcomes

4. A coin is tossed, and a 6-sided number cube is rolled.

   _____ outcomes

5. The names of 6 people are written on slips of paper. Two slips are drawn with replacement.

   _____ outcomes

6. A spinner is divided in half. The spinner is spun, and a coin is tossed.

   _____ outcomes

7. A spinner is divided in 5 equal parts. The spinner is spun, and a 6-sided number cube is rolled.

   _____ outcomes

8. Two 4-part spinners are spun. One spinner has the numbers 1–4 labeled; the other has the letters A–D labeled.

   _____ outcomes

9. A spinner has 3 equal parts colored red, blue, and yellow. The spinner is spun, and a coin is tossed.

   _____ outcomes

10. The numbers 1 through 10 are written on slips of paper. Two slips are drawn with replacement.

    _____ outcomes

11. **Critical Thinking** Three 6-sided number cubes are rolled. In how many of the outcomes is the sum of the 3 numbers even? _____

Problem Solving/Critical Thinking

4.4 Solving a Simpler Problem

With some data sets, you may be able to simplify the calculation of the mean. Instead of working directly from the actual values, you can work with easier numbers. This strategy requires that you first make an estimate for the mean. If you practice this strategy you may eventually be able to do the calculations mentally.

◆ **Example**
   Find the mean of 92, 87, 90, 93, and 87.

◆ **Solution**
   First, estimate a mean. All of the numbers are close to 90, so you might use 90 as the estimate.
   Subtract the estimated mean, 90, from each of the numbers in the data set. Indicate differences as positive or negative numbers, using two columns.

◆ **Computation**
   estimated mean: 90

92 − 90 =	+2	
87 − 90 =		−3
90 − 90	0	
93 − 90	+3	
87 − 90 =		−3

Add the positive and negative differences.   +5   −6
Add the sums.   5 + (−6) = −1
Divide the total by 5, the number of items in the data set.   (−1) ÷ 5 = −0.2
Adjust your estimate, 90, downward by 0.2.   90 − 0.2 = 89.8

You have found the actual mean of the data.   actual mean: 89.8

Use the example to solve these problems.

1. Confirm that the strategy results in the actual mean by adding the five numbers and then dividing by 5. Show your work.

   _____

2. Use an estimate different from 90 for the mean. Show that the strategy still results in the same actual mean.

   _____

Use the strategy from the example to find the actual mean for each data set. Round answers to the nearest tenth.

Data set	Estimated mean	Actual mean
3. 76, 74, 77, 73, 78	_____	_____
4. 50, 48, 52, 49, 51, 55, 50, 53	_____	_____
5. 30, 41, 32, 36, 37, 38, 34, 40, 32, 42, 36, 34	_____	_____
6. 83, 87, 90, 90, 81, 85, 86, 79, 81, 77, 79, 84, 76, 81	_____	_____
7. 128, 123, 125, 124, 127, 128, 122, 126, 125, 120	_____	_____

8. **Critical Thinking** A data set contains three items. Would you be likely to use the strategy explained on this page for such a small data set? Explain why or why not.

_____

Problem Solving/Critical Thinking
4.5 Planning a Graph

Some data sets are more appropriately shown on bar graphs; other data sets are represented better with line graphs.

Uses for bar graphs	Uses for line graphs
If a set of separate quantities are being compared, try a bar graph. A comparison of heights of a group of people would make a good bar graph.	If something is changing over time, try a line graph. A line graph might show the change in height of a plant grown for a science experiment.

Once you have chosen the kind of graph to make, the next step is to decide on the scales of the axes. Either quantity can be shown on the horizontal scale, and a scale can be numbered in multiples of any number.

Complete this chart that has been started for you. You may want to make sketches of the graphs to help you in your decision-making.

	Data	Graph type	Horizontal axis	Vertical axis
1.	heights of 10 basketball players from 5 feet 10 inches to 7 feet 3 inches	bar	names of the 10 people	
2.	weekly sales of a company over a 2-year period from $23,000 through $68,000	line		sales in dollar amounts numbered in multiples of $5000
3.	a survey of 2000 households asking how many cars in each household		number of cars from 0 through 10	
4.	survey of 100 people asking their favorite kind of movie			
5.	average number of airplane trips taken in the last month by people in different income brackets			
6.	monthly balance in a person's savings account over the last year			
7.	heights in centimeters of two groups of plants: one grown with fertilizer and one grown without			
8.	the temperature at noon on the first day of each week for a period of 2 months			

9. **Critical Thinking** Describe how the overall size of a graph, such as a space 3 inches wide and 2 inches deep, can determine the choice of scales.

_____

Problem Solving/Critical Thinking

4.6 *Comparing Data Representations*

There is almost always more than one way to represent a set of data. On this page, you will compare the usefulness of a stem-and-leaf plot to that of a histogram.

key | 1 | 7 = 17 |

Stem	Leaf
1	7 8 9
2	3 5 6 6 7
3	2 4 5 6 8 8 9
4	0 1 3 4 6 9
5	0 2

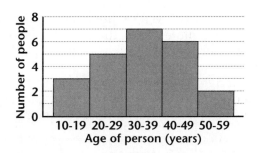

Complete this chart. Use the last column to record whether the question can be answered using the histogram, the stem-and-leaf plot, neither, or both. If both graphs can be used, note which you think is the easier to use.

	Question	Answer	Graph(s) used
1.	How many people are in the data set?		
2.	How old is the oldest person?		
3.	How many people are younger than 10?		
4.	How many people are 50 or older?		
5.	How many people are married?		
6.	What is the range of the data?		
7.	Which interval has the lowest frequency?		
8.	How many people own a car?		
9.	How many people are old enough to vote?		
10.	Are there more people under 30 or over 39?		
11.	What is the median of the data?		
12.	What is (are) the mode of the data?		

13. **Critical Thinking** Do you prefer the stem-and-leaf plot or the histogram? Give reasons for your preference.

_____

_____

_____

Problem Solving/Critical Thinking
5.1 Using Imaginary Lines

Shown at right is the graph of an equation. To find out whether the graph represents a function, you can use the vertical-line test. It states the following:

If there is at least one vertical line that cuts through a graph in two or more points, then the graph does not represent a function.

In the diagram at right, notice that l_1, l_2, and l_4 each intersect the graph in one point. However, l_3 intersects the graph in three points. Thus, this graph does not represent a function.

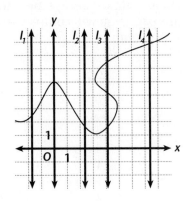

In Exercises 1–6, tell whether the graph represents a function. If the graph does not represent a function, give an explanation.

1. _____

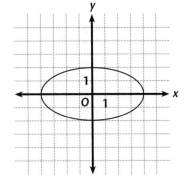

2. _____

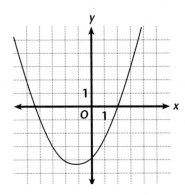

3. _____

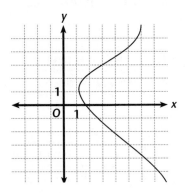

4. _____

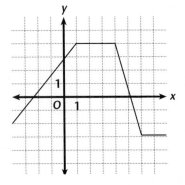

5. _____

6. _____

7. Explain why the graph described at right does not represent a function.

_____

On a coordinate grid, a continuous curve moves to the right, then goes straight up, then doubles back to the left, and finally moves up and to the right.

Problem Solving/Critical Thinking

5.2 *Sketching Slope from Two Points*

If you are given the coordinates of two points, you can find the slope of the line through those points by subtracting the coordinates. Making a sketch can help you check your computation and show whether the slope is positive or negative.

◆ **Example**

Sketch the relative positions of the points $A(9, 6)$ and $B(2, 7)$.

◆ **Solution**

Step 1	Step 2	Step 3
Make a dot for either point. Label the coordinates of the point. We'll start with point A. (9, 6) ● A	Compare the *x*-coordinates. Since 2 is less than 9, point *B* is to the left of point *A*. Compare the *y*-coordinates. Since 7 is greater than 6, point *B* is above point *A*.	Put point *B* in the correct relative position to point *A*. 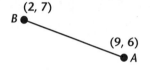

Describe how point *B* relates to point *A* by circling the appropriate words.

1. $A(-3, 5), B(-1, 2)$
Point *B* is to the left/right of point *A*.
Point *B* is above/below point *A*.

2. $A(5, -3), B(2, -1)$
Point *B* is to the left/right of point *A*.
Point *B* is above/below point *A*.

3. $A(-1, -2), B(3, 1)$
Point *B* is to the left/right of point *A*.
Point *B* is above/below point *A*.

4. $A(0, -3), B(7, -6)$
Point *B* is to the left/right of point *A*.
Point *B* is above/below point *A*.

5. $A(6, 4), B(8, -2)$
Point *B* is to the left/right of point *A*.
Point *B* is above/below point *A*.

6. $A(5, 3), B(-1, 6)$
Point *B* is to the left/right of point *A*.
Point *B* is above/below point *A*.

7. $A(2, -2), B(-6, 0)$
Point *B* is to the left/right of point *A*.
Point *B* is above/below point *A*.

8. $A(-2, 2), B(-6, 4)$
Point *B* is to the left/right of point *A*.
Point *B* is above/below point *A*.

Sketch the relative positions of each pair of points. Then describe the slope as positive or negative.

9. $A(-1, 2), B(-6, 5)$ **10.** $A(0, -4), B(2, -6)$ **11.** $A(1, -2), B(4, 1)$

slope: _____ slope: _____ slope: _____

12. Critical Thinking Describe the line for which

a. two points have the same *x*-coordinate. _____

b. two points have the same *y*-coordinate. _____

Problem Solving/Critical Thinking

5.3 *Missing Information: Changing Units in Rates*

When you are solving problems that involve rates, you may need to change units. To do this, you will need a conversion equation, such as 1 foot = 12 inches. This kind of information can be found in textbooks and in large dictionaries.

◆ **Example**
Art ran 400 meters in 3 minutes. Find his speed in miles per hour.

◆ **Solution**
Multiply by ratios and show the conversion factors.

speed = 400 meters in 3 minutes

$$= \frac{400 \text{ meters}}{3 \text{ minutes}} \cdot \frac{1 \text{ mile}}{1609.344 \text{ meters}} \quad \longleftarrow \text{ Convert meters to miles.}$$

$$= \frac{0.2485 \text{ miles}}{3 \text{ minutes}}$$

$$= \frac{0.2485 \text{ miles}}{3 \text{ minutes}} \cdot \frac{60 \text{ minutes}}{1 \text{ hour}} \quad \longleftarrow \text{ Convert minutes to hours.}$$

$$= 4.97 \text{ miles per hour}$$

Describe the purpose of each statement.

1. 14 liters per minute $= \dfrac{14 \text{ liters}}{1 \text{ minute}} \cdot \dfrac{1 \text{ gallon}}{3.785 \text{ liters}} \cdot \dfrac{60 \text{ minutes}}{1 \text{ hour}} = \boxed{?}$ _____

2. 5 ounces per day $= \dfrac{5 \text{ ounces}}{1 \text{ day}} \cdot \dfrac{1 \text{ pound}}{16 \text{ ounces}} \cdot \dfrac{7 \text{ days}}{1 \text{ week}} = \boxed{?}$ _____

3. 8.27 centimeters per second $= \dfrac{8.27 \text{ cm}}{1 \text{ second}} \cdot \dfrac{60 \text{ seconds}}{1 \text{ minute}} \cdot \dfrac{1 \text{ foot}}{30.48 \text{ cm}} = \boxed{?}$ _____

4. 12 pounds per minute $= \dfrac{12 \text{ pounds}}{1 \text{ minute}} \cdot \dfrac{373.24 \text{ grams}}{1 \text{ pound}} \cdot \dfrac{1 \text{ minute}}{60 \text{ seconds}} = \boxed{?}$ _____

5. 45 kilometers per week $= \dfrac{45 \text{ km}}{1 \text{ week}} \cdot \dfrac{1 \text{ week}}{168 \text{ hours}} \cdot \dfrac{1000 \text{ m}}{1 \text{ km}} = \boxed{?}$ _____

6. 16 millimeters per second $= \dfrac{16 \text{ mm}}{1 \text{ second}} \cdot \dfrac{1 \text{ inch}}{25.4 \text{ mm}} \cdot \dfrac{60 \text{ seconds}}{1 \text{ minute}} = \boxed{?}$ _____

Use information on this page to convert each rate to the desired units.

7. Change $2\frac{1}{2}$ gallons per minute to liters per second. _____

8. Change 75 centimeters per second to feet per hour. _____

9. Change 50 ounces per second to pounds per minute. _____

10. Change 650 grams per minute to pounds per hour. _____

11. **Critical Thinking** A dictionary says there are 16 pints in 1 peck. It also says that 1 pint = 0.0625 pecks. Which of these conversion equations can be used to convert pints into pecks? Show how you would do this kind of conversion.

_____

Problem Solving/Critical Thinking
5.4 Estimating Slope

Making a visual estimate of the slope of a line is a useful skill. You can use it to check your work when drawing graphs of linear equations, or when writing the equation for a given graph.

Graphs with positive slopes

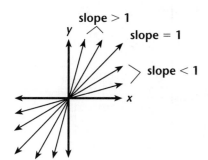

Graphs with negative slopes

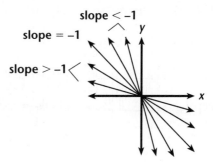

Notice that a line with a slope of 1 or −1 is a diagonal of a square.

Caution: When making a visual estimate of slope from a graph, make sure that the scales on the two axes are numbered with the same increments. Otherwise, a line with a slope of 1 will not have a slope that looks like a 45° diagonal line.

Describe the slope of each line. First, write whether the slope is positive or negative. Then compare the slope to 1 or −1.

1.

_____

_____

2.

_____

_____

3.

_____

_____

4.

_____

_____

5.

_____

_____

6.

_____

_____

7.

_____

_____

8.

_____

_____

9. **Critical Thinking** The absolute value of the slope of a certain line is a fraction between 0 and 1. Describe what this line looks like.

_____

_____

Problem Solving/Critical Thinking
5.5 *Using Mental Math to Find Intercepts*

In a linear equation, when x equals 0, the value of y is the y-intercept. When $y = 0$, the value of x is the x-intercept. Here is a way to use mental math to find the two intercepts and, therefore, graph a linear equation in x and y.

x-intercept: Put your finger over the y-term and mentally solve for x.

y-intercept: Put your finger over the x-term and mentally solve for y.

Use mental math to find the intercepts for the graph of each equation.

1. $2y + 3 = -4x$

y-intercept: _____

x-intercept: _____

2. $-1 - x = 5y$

y-intercept: _____

x-intercept: _____

3. $x + 3y - 6 = 0$

y-intercept: _____

x-intercept: _____

4. $2x - y = 4$

y-intercept: _____

x-intercept: _____

5. $-3y - 1 = 6x$

y-intercept: _____

x-intercept: _____

6. $8 + y = -4x$

y-intercept: _____

x-intercept: _____

7. $6 = 3x + y$

y-intercept: _____

x-intercept: _____

8. $3x + 2 = -2y$

y-intercept: _____

x-intercept: _____

9. $5 - x + y = 0$

y-intercept: _____

x-intercept: _____

10. $-4x + 6 = 2y$

y-intercept: _____

x-intercept: _____

11. $4y = 8 - 3x$

y-intercept: _____

x-intercept: _____

12. $-2 = 3y + x$

y-intercept: _____

x-intercept: _____

Graph each equation by finding the two intercepts.

13. $3x - 6 = 2y$

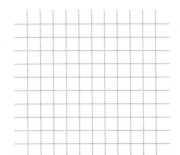

14. $y + 4x + 4 = 0$

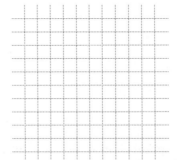

15. $-2 = 2x - y$

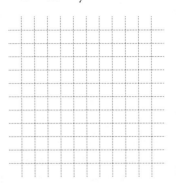

16. Critical Thinking Describe linear equations that do not have both an x- and a y-intercept.

_____

Problem Solving/Critical Thinking
5.6 *Discarding Unnecessary Information*

For some problems, you are given more information than you need. When this is the case, focus on only the key ideas or facts.

◆ **Example**
   Determine if these two lines are parallel or perpendicular. $2x + 4 = 3y$ and $4y - 11 = -6x$

◆ **Solution**
   You need only the slopes of the lines. Ignore the constant terms.
   The two equations in this example become $2x = 3y$ and $4y = -6x$.
   Now solve for y in each equation. You will get $y = \frac{2}{3}x$ and $y = \frac{-3}{2}x$.

   The slopes are negative reciprocals, so the lines are pependicular.

Find the slope of each line. Then tell whether the pair of lines is parallel, perpendicular, or neither.

1. $1 = 10x - 5y$ $2x - y = 4$

    slope: _____   slope: _____

    _____

2. $-1 = 5x - y$ $-1 - x = 5y$

    slope: _____   slope: _____

    _____

3. $4x - 2y = -3$ $2y + 3 = -4x$

    slope: _____   slope: _____

    _____

4. $5 - x + y = 0$ $2x - 5 + 2y = 0$

    slope: _____   slope: _____

    _____

5. $x + 3y - 6 = 0$ $-2y + 6 = 3x$

    slope: _____   slope: _____

    _____

6. $12x = 5 - 4y$ $6 = 3x + y$

    slope: _____   slope: _____

    _____

7. $4y = 8 - 3x$ $6x + 8y = 1$

    slope: _____   slope: _____

    _____

8. $-3y - 1 = 6x$ $2y - x + 3 = 0$

    slope: _____   slope: _____

    _____

9. $3 = x - 4y$ $8 + y = -4x$

    slope: _____   slope: _____

    _____

10. $2x + 3y = -1$ $3x + 2 = -2y$

    slope: _____   slope: _____

    _____

11. **Critical Thinking** What happens if one of the lines is horizontal or vertical? What do you do then?

    _____

    _____

Problem Solving/Critical Thinking

6.1 Predicting Solutions to Inequalities

The solution to an inequality is usually a large set of numbers. The set can include all positive numbers, all negative numbers, or some of both. By thinking about an inequality before you solve it, you can predict the kind of solution set that you will get.

Examples	**Solutions**
◆ $n - 2.3 > 1.2$	◆ Add 2.3 to each side so that all the solutions will be positive.
◆ $4 + n < -2$	◆ Subtract 4 from each side so that all the solutions will be negative.
◆ $6 + n > 3$	◆ Subtract 6 from each side. Some solutions will be positive and some will be negative.

Describe the solution set for each inequality as *all positive, all negative,* or *some of both.*

1. $f + 12 > -2$

2. $r - 7 < 10$

_____

3. $m + 1.5 < 0.5$

4. $3.4 + h > -12$

_____

5. $d - 4 > -2$

6. $-9 + p < 2$

_____

Without solving the inequality, check the one box that best describes the solution.

7. $a - 7 < 9$
The solutions will all be
☐ greater than 2.
☐ less than 2.
☐ greater than 16.
☐ less than 16.

8. $n + 2.3 > 8$
The solutions will all be
☐ greater than 10.3.
☐ less than 10.3.
☐ greater than 5.7.
☐ less than 5.7.

9. $1 < y - 3$
The solutions will all be
☐ greater than 4.
☐ less than 4.
☐ greater than -2.
☐ less than -2.

10. $1.2 > g + 2.5$
The solutions will all be
☐ greater than 3.6.
☐ less than 1.3.
☐ greater than -3.6.
☐ less than -1.3.

11. Critical Thinking Explain why the solution sets of most inequalities contain both integers and fractions.

_____

NAME _____ CLASS _____ DATE _____

Problem Solving/Critical Thinking

6.2 *Solving a Related Problem*

Suppose that you are good at solving equations but you find inequalities confusing. Here is a strategy you can use to solve inequalities.

◆ **Example**   Solve $7 - 3x > 16$.

◆ **Solution**

> **Step 1**   For the time being, replace the inequality sign with an equal sign. Then solve this related equation.
>
> $$7 - 3x = 16$$
> $$7 - 3x - 7 = 16 - 7$$
> $$-3x = 9$$
> $$-3x \div -3 = 9 \div -3$$
> $$x = -3$$
>
> **Step 2**   Now decide whether to replace the $=$ with $>$ or $<$. Find any number that makes the original inequality true, for example, $x = -10$. Compare the two possible choices, $x > -3$ and $x < -3$. For $x = -10$, the inequality $x < -3$ is true. Therefore, the solution to the original inequality is $x < -3$.

The related equation and its solution are given for each inequality. Use this information to write the solution for the inequality.

1. $3 - x > 4$
    If $3 - x = 4$, $x = -1$.

_____

2. $7 < -2 + x$
    If $7 = -2 + x$, $x = 9$.

_____

3. $-12 < x - 5$
    If $-12 = x - 5$, $x = -7$.

_____

4. $4x + 3 > -2$
    If $4x + 3 = -2$, $x = -1.25$.

_____

5. $5 - 2x < -13$
    If $5 - 2x = -13$, $x = 9$.

_____

6. $-3 - 2x < 9$
    If $-3 - 2x = 9$, $x = -6$.

_____

7. $0.2x + 1.5 < -2.5$
    If $0.2x + 1.5 = -2.5$, $x = -20$.

_____

8. $1.2 > 5x + 7.7$
    If $1.2 = 5x + 7.7$, $x = -1.3$.

_____

9. $0.9 > 3.1 - 0.1x$
    If $0.9 = 3.1 - 0.1x$, $x = 22$

_____

10. $\frac{x}{5} - 2 > -12$
    If $\frac{x}{5} - 2 = -12$, $x = -50$.

_____

11. $3 < 4 - \frac{x}{2}$
    If $3 = 4 - \frac{x}{2}$, $x = 2$.

_____

12. $-\frac{1}{2} < \frac{3x}{4} - \frac{5}{8}$
    If $-\frac{1}{2} = \frac{3x}{4} - \frac{5}{8}$, $x = \frac{1}{6}$.

_____

13. **Critical Thinking**   Describe how to adapt the strategy on this page for an inequality that uses $\leq$ or $\geq$ rather than $<$ or $>$.

_____

_____

Problem Solving/Critical Thinking

6.3 Using Equations to Solve Compound Inequalities

Shown at right is the solution to the inequality $-3 \le 2x - 1 \le 7$. Shown below is the solution set on a number line. Notice that the two endpoints of the interval that give the solution are the solutions to the equations $-3 = 2x - 1$ and $2x - 1 = 7$.

$$-3 \le 2x - 1 \le 7$$
$$-2 \le 2x \le 8$$
$$-1 \le x \le 4$$

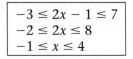

This reasoning suggests the following:

If you have a compound inequality in x that involves conjunction, then you can solve it by solving a pair of equations, plotting the two solutions on a number line, and joining them with a line segment.

In Exercises 1–6, solve the pair of related equations that give the endpoints of the solution to the given inequality. Graph the solutions to the equations on each number line.

1. $-3 \le 2x + 5 < 10$ _____

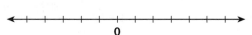

2. $0 < 5x + 5 < 15$ _____

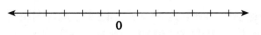

3. $-12 < 3x < 15$ _____

4. $20 \le 5x + 5 < 30$ _____

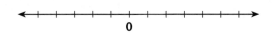

5. $-12 < 7x + 2 \le 16$ _____

6. $0 \le -2x + 1 \le 10$ _____

If you have a compound inequality in which the inequalities are joined by *or*, you can also use related equations to help determine the solution to the given compound inequality.

In Exercises 7 and 8, solve the pair of related equations that give the endpoints of the solution to the given inequality. Graph the solutions to the equations on each number line.

7. $2x - 9 < -12$ or $2x + 5 \ge 12$ _____

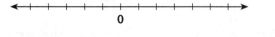

8. $-3x + 6 > 0$ or $3x + 2 > 11$ _____

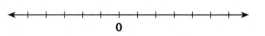

9. **Critical Thinking** Write one inequality involving *and* and one inequality involving *or* that together give the diagram at right as the solution.

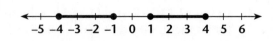

_____

Problem Solving/Critical Thinking

6.4 *Graphing Translations of Absolute-Value Functions*

Functions of the form $y = |x - h| + k$ are translations of the basic absolute-value function. The graphs of these functions have the same shape as the graph of $y = |x|$, but the tip of the V is located at (h, k) because $|x - h| = 0$ when $x = h$.

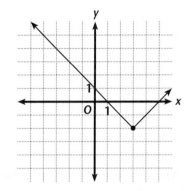

♦ **Example** Graph the function $y = |x - 3| - 2$.

This equation is of the form $y = |x - h| + k$, where $h = 3$ and $k = -2$. The tip of the V is located at $(3, -2)$.

Graph each function.

1. $y = |x - 1| - 4$

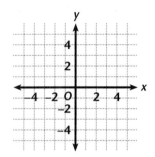

2. $y = |x| + 3$

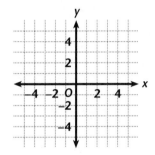

3. $y = |x - 2|$

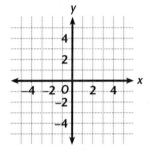

4. $y = |x - 3| - 1$

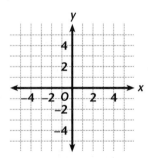

5. $y = |x| - 5$

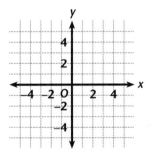

6. $y = |x + 1| - 2$

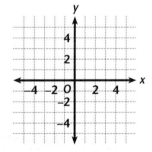

7. $y = |x + 3| + 2$

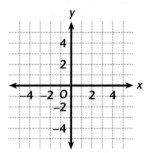

8. $y = |x - 4| - 3$

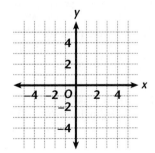

9. $y = |x + 2| + 1$

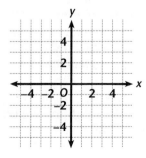

10. Critical Thinking Explain why h is subtracted but k is added in $y = |x - h| + k$.

_____

_____

Problem Solving/Critical Thinking

6.5 Applying a Definition

To solve an equation such as $|x - 3| \geq 6$, you can use the definition of absolute value. Here are some implications of this definition.

$$|x| = a \longrightarrow \qquad +x = a \qquad -x = a$$
$$|x| = 6 \longrightarrow \qquad +x = 6 \qquad -x = 6$$
$$|x - 3| = 6 \longrightarrow \quad +(x - 3) = 6 \quad -(x - 3) = 6$$
$$|x - 3| \geq 6 \longrightarrow \quad +(x - 3) \geq 6 \quad -(x - 3) \geq 6$$

Write two statements that are equivalent to each equation or inequality.

1. $|2x + 5| = 3$

_____

_____

2. $|7 - x| = 2$

_____

_____

3. $|-3x - 5| = 8$

_____

_____

4. $|7 - 4x| > 1$

_____

_____

5. $|3 + 2x| \leq 5$

_____

_____

6. $|-x + 3| \geq 6$

_____

_____

7. $5 = |5x + 2|$

_____

_____

8. $4 < |6 - 2x|$

_____

_____

9. $3 \geq |4x + 1|$

_____

_____

Critical Thinking The absolute value of a quantity must always be positive. Use this fact to describe the solutions for each inequality. Check one box for each problem.

10. $|3x + 6| < -3$
The solutions will be
☐ no real numbers.
☐ some real numbers.
☐ all real numbers.

11. $|7 - 2x| > 2$
The solutions will be
☐ no real numbers.
☐ some real numbers.
☐ all real numbers.

12. $|4x + 1| > -1$
The solutions will be
☐ no real numbers.
☐ some real numbers.
☐ all real numbers.

13. $|-3 - 2x| \leq 4$
The solutions will be
☐ no real numbers.
☐ some real numbers.
☐ all real numbers.

14. $|-x + 5| \geq -5$
The solutions will be
☐ no real numbers.
☐ some real numbers.
☐ all real numbers.

15. $|x - 8| \leq -8$
The solutions will be
☐ no real numbers.
☐ some real numbers.
☐ all real numbers.

Problem Solving/Critical Thinking
7.1 Defining Variables

When you set up a system of equations to solve a problem, you need at least two variables. Often the letters x and y are used for these variables. But you can use any letters you choose. It may be useful to choose letters that help you remember what the variables mean; for example, d for distance, t for time, or c for cost.

Match each problem to a pair of variables. Write the letter of the best match in the line next to the problem number. Do not solve the given problem. (One pair of variables will not be used.)

_____ **1.** The difference between two numbers is 18. Twice the smaller number plus 3 times the larger is 74. What are the numbers?

_____ **2.** George has $1.30 in dimes and nickels. If there are 19 coins in all, how many dimes does George have?

_____ **3.** There are two boat rental shops at a lake. Best Boats charges $25 per hour plus a deposit of $50. Hal's Boat Rental charges $30 per hour plus a deposit of $15. Which is the least expensive choice?

_____ **4.** An airplane starts at 30,000 feet and descends at 1500 feet per minute. Another airplane takes off and ascends at 1800 feet per minute. How many minutes will it take for the planes to be at the same altitude?

_____ **5.** The perimeter of a rectangle is 400 meters. The length is 40 meters more than the width. Find the length and the width.

_____ **6.** The highest possible score on a history test is 150 points. There are 18 questions in all, 12 multiple-choice questions worth 5 points each and 6 essay questions worth 15 points each. How many questions of each type are on this test?

_____ **7.** Ed is two-thirds as old as Al. In 7 years, Ed will be three-fourths as old as Al. How old is each person now?

_____ **8.** Two cars leave the same town at the same time going in the same direction. One car travels 30 miles per hour and the other travels 46 miles per hour. In how many hours will the cars be 72 miles apart?

A. $h =$ number of hours
$c =$ total cost

B. $m =$ number of multiple-choice questions
$e =$ number of essay questions

C. $d =$ number of dimes
$n =$ number of nickels

D. $h =$ number of hours
$m =$ number of miles

E. $s =$ distance traveled by slower car
$f =$ distance traveled by faster car

F. $a =$ altitude in feet
$m =$ number of minutes

G. $s =$ smaller number
$g =$ greater number

H. $W =$ width
$L =$ length

I. $E =$ Ed's age now
$A =$ Al's age now

9. Critical Thinking The sum of the digits of a 3-digit number is 9. If the digits are reversed, the number increases by 495. The sum of the tens and hundreds digits is one-half the ones digit. Find this 3-digit number. _____

Problem Solving/Critical Thinking

7.2 *Adapting the Substitution Method*

It may appear that the system at the right is one that you cannot solve by using substitution. However, you do not need to have one variable with a coefficient of 1 in order to solve a system of equations by the substitution method.

$$\begin{cases} 2x + 3y = 21 \\ -3x - 6y = -24 \end{cases}$$

◆ **Example** Solve $\begin{cases} 2x + 3y = 21 \\ -3x - 6y = -24 \end{cases}$.

◆ **Solution** Focus on the *y*-terms. Solve the first equation for $3y$. In the second equation write $6y$ as $2(3y)$. Now you can substitute for $3y$.

$$\begin{cases} 2x + 3y = 21 \\ -3x - 6y = -24 \end{cases} \longrightarrow \begin{cases} 3y = 21 - 2x \\ -3x - 2(3y) = -24 \end{cases} \longrightarrow -3x - 2(21 - 2x) = -24$$

Now you have achieved your goal of getting one equation in one variable.

The solution to the system is $x = 18$ and $y = -5$.

In Exercises 1–5:
Step 1: Solve one equation for the expression in bold.
    Write the other equation using a multiple of the expression in bold.
Step 2: Write a single equation involving the variable that is not bold.
Step 3: Solve the system of equations.

	System	Step 1	Step 2	Step 3
1.	$\begin{cases} 9 - \mathbf{2x} + 3y = 0 \\ 4x - 51 = 5y \end{cases}$			
2.	$\begin{cases} 9y - 5 = -2x \\ 5x + \mathbf{3y} = -7 \end{cases}$			
3.	$\begin{cases} 15x - 2y + 75 = 0 \\ -\mathbf{5x} + 3y = 25 \end{cases}$			
4.	$\begin{cases} 3x - \mathbf{4y} = 6 \\ 8y - 7x = -18 \end{cases}$			
5.	$\begin{cases} 21 + 10x = 3y \\ 2y - \mathbf{5x} = 9 \end{cases}$			

6. **Critical Thinking** Describe the characteristics of a system of equations in which you could use this method of solution.

_____

_____

_____

Problem Solving/Critical Thinking
7.3 Organizing Work in a Chart

The elimination method of solving a system of equations involves quite a
number of steps. Here is a way to organize part of the work in a chart.

Step 1	Step 2	Step 3	Step 4
Put both equations in the form $Ax + By = C$.	Choose the factors that you will use to multiply the equations.	Multiply one or both equations so that you can eliminate a variable.	Add the equations to eliminate one of the variables.
$-2x - 5y = -1 \longrightarrow$ $3x + 2y = -4 \longrightarrow$	$2 \cdot [-2x - 5y = -1] \longrightarrow$ $5 \cdot [3x + 2y = -4] \longrightarrow$	$-4x - 10y = -2 \longrightarrow$ $15x + 10y = -20 \longrightarrow$	$-4x - 10y = -2$ $15x + 10y = -20$
			$11x \qquad = -22$ $x = -2$

1. Finish the example problem by finding y. _____

2. Find the solution to the system that contains $2x - 9 + 5y = 0$ and $4 + 2y = 3x$.

Step 1	Step 2	Step 3	Step 4

solution: _____

3. Find the solution to the system that contains $4y + 16 = 3x$ and $5x = 14 - 6y$.

Step 1	Step 2	Step 3	Step 4

solution: _____

Make a chart to help you solve each system.

4. $\begin{cases} 5x = 2y \\ 3y - 11 - 2x = 0 \end{cases}$
5. $\begin{cases} y + 8 = -x \\ 2x + 1 = y \end{cases}$
6. $\begin{cases} 2x = 3y - 1 \\ 4y = 24 - 3x \end{cases}$

_____ _____ _____

7. **Critical Thinking** On a separate sheet of paper, create a different
   type of chart that someone could use to solve this kind of problem. Put
   the numbers of the steps in the left column instead of in the top row.

NAME _____ CLASS _____ DATE _____

Problem Solving/Critical Thinking
7.4 Predicting Solutions to Systems

There are three possibilities for the solution to a system of equations. To
decide which possibility applies, put both equations in the form
$Ax + By = C$. Then look at the relationship between the equations.

	Case 1	Case 2	Case 3
Example	$-4x + 8y = 14$ $-2y + 4y = 7$	$-4x + 8y = 14$ $-2y + 4y = 10$	$-4x + 8y = 14$ $-2y + 3y = 7$
Observation	One equation is a multiple of the other.	The left side of one equation is a multiple of the other equation.	Neither case 1 nor case 2 is true.
What the graph will look like	The lines coincide.	The lines are parallel.	The lines intersect in a single point.
Number of Solutions	infinitely many	none	one

**Put each equation in the form $Ax + By = C$. Then predict the
number of solutions.**

1. $\begin{cases} 3x = 2 - y \\ 5y + 10 = 2x \end{cases}$

2. $\begin{cases} 3y - 2 = 4x \\ 12x - 9y + 6 = 0 \end{cases}$

3. $\begin{cases} 3y - 2x = 1 \\ 2x = 6y - 1 \end{cases}$

4. $\begin{cases} 25y + 5 - 10x = 0 \\ 2x - 1 = 5y \end{cases}$

5. $\begin{cases} 4y + 3x + 8 \\ 5x + 2 = y \end{cases}$

6. $\begin{cases} 6x = 10 + 9y \\ 3y + 5 = 2x \end{cases}$

7. $\begin{cases} x = 7y + 2 \\ 21y + 6 = 3x \end{cases}$

8. $\begin{cases} y - 2x = -4 \\ 8x + 1 = 4y \end{cases}$

9. $\begin{cases} 1 + 4x = 2y \\ y - 2 = 8x \end{cases}$

10. **Critical Thinking** Two equations in a system are both in the form
$y = mx + b$. Describe how to predict the number of solutions to this
system.

Problem Solving/Critical Thinking

7.5 Classifying the Solution

Before you start to solve a problem, think about the nature of what you are trying to find. The problem might be asking for a number, such as 23; an ordered pair, such as $(-2, 5)$; a quantity, such as 4 square feet; or a phrase, such as *no solution*.

What is each problem asking you to find? Choose your answers from this box.

a number	a word or phrase	a set of ordered pairs
a quantity	an ordered pair	an equation

1. One number is 5 times another number. Their sum is 102. What is the smaller number?

   _____

2. What is the solution of this system?
   $2x + 3y = 1 \qquad -y + 3x - 5 = 0$

   _____

3. What is the solution of $2x - 5 = -1$?

   _____

4. What is 120 percent of $500?

   _____

5. Is this system of equations consistent or inconsistent?
   $2x - y + 3 = 0 \qquad 1 + 2y = -4x$

   _____

6. Marge is 4 times as old as Susan. In 8 years, Marge will be twice as old as Susan. How old is Susan now?

   _____

7. What is the slope-intercept form of the equation $4y - 3x = -1$?

   _____

8. What is a line parallel to $2y - x = 3$ that contains the point $P(2, -1)$?

   _____

9. Describe the points that are on the line that connects the points $A(2, -4)$ and $B(0, 3)$.

   _____

10. What value of x makes the expression $2x - 5$ equal to 0?

    _____

11. What is the solution of $2y > 3 - x$?

    _____

12. What is the solution of $3x - y = -2$?

    _____

13. Are these lines parallel?
    $x + 6 = -3y \qquad 3x - 6y = 1$

    _____

14. What is the y-intercept of the graph of $2x + y = 4$?

    _____

15. **Critical Thinking** What kinds of math problems have solutions that are drawings rather than numbers or words? _____

Problem Solving/Critical Thinking
7.6 Remembering the Main Idea

Imagine this situation. For a complicated word problem, a student sets up a system of equations. After a lengthy series of steps, the student gets an answer such as $x = 32.5$. At this point, the student has forgotten what he or she was looking for and doesn't have a clue what x means! The problems on this page will help you practice keeping the main idea of a problem in mind.

What does each problem ask you to find? Check the best answer.

1. A chemist has one solution that is 75% chlorine and another that is 40% chlorine. How much of each is needed to make a 100 liter solution that is 50% chlorine?

☐ volume of each solution in liters
☐ volume of chlorine in each solution
☐ total amount of chlorine in each solution
☐ amount of chlorine in the 100 L solution
☐ percent of chlorine in both solutions

2. In an 8-kilometer race, a runner ran part of the distance and walked part of the distance. Her walking speed is 7 kilometers per hour and her running speed is 10 kilometers per hour. How far did she run?

☐ the time she spent walking
☐ the time she spent running
☐ the number of kilometers she ran
☐ the number of kilometers she walked
☐ the time it took her to finish the race

3. A train leaves Los Angeles for Boston, traveling 72 kilometers per hour. Three hours later a second train leaves on a parallel track and travels at 120 kilometers per hour. When will the second train catch up to the first train?

☐ distance the second train travels
☐ time the second train leaves
☐ number of hours until the trains are the same distance from Los Angeles
☐ time it takes the trains to get to Boston

4. There were 578 tickets sold for a school concert. Tickets were $2 for adults and $1.50 for children. How much was collected from the adults?

☐ number of children's tickets sold
☐ dollars collected for children's tickets
☐ total number of dollars collected
☐ number of adults' tickets sold
☐ dollars collected for adults' tickets

5. Hal saved quarters and dimes for parking meters. He had 103 coins worth a total of $15.25. How many dimes did he have?

☐ value of the dimes
☐ value of the quarters
☐ number of dimes that Hal had
☐ number of quarters that Hal had
☐ number of coins that Hal had

6. A 30–gallon barrel of milk is 4% butterfat. How much skim milk (no butterfat) should be mixed to make milk that is 1.5% butterfat?

☐ gallons of milk
☐ percent of milk
☐ gallons of butterfat
☐ gallons of skim milk
☐ percent of skim milk

7. Critical Thinking Describe how you would check the answer to Exercise 6.

_____

_____

Problem Solving/Critical Thinking

8.1 Diagramming Products of Monomials

Making a diagram can help you find the product of monomials. The diagram lets you break the problem into smaller parts.

◆ **Example**   Simplify the product $(6a^2b^3)(7b^3c^2)$.

◆ **Step 1**   Make the diagram. Use one box for the coefficient and one box for each exponent in the product.

$(6a^2b^3)(7b^3c^2) \longrightarrow \square\, a^{\square}\, b^{\square}\, c^{\square}$

◆ **Step 2**   Multiply the coefficients: $6 \cdot 7 = 42$.

$(6a^2b^3)(7b^3c^2) \longrightarrow \boxed{42}\, a^{\square}\, b^{\square}\, c^{\square}$

◆ **Step 3**   Find the first exponent: a^2.

$(6a^2b^3)(7b^3c^2) \longrightarrow \boxed{42}\, a^{\boxed{2}}\, b^{\square}\, c^{\square}$

◆ **Step 4**   Add to find the second exponent: $b^3 \cdot b^3 = b^6$.

$(6a^2b^3)(7b^3c^2) \longrightarrow \boxed{42}\, a^{\boxed{2}}\, b^{\boxed{6}}\, c^{\square}$

◆ **Step 5**   Find the third exponent: c^2.

$(6a^2b^3)(7b^3c^2) \longrightarrow \boxed{42}\, a^{\boxed{2}}\, b^{\boxed{6}}\, c^{\boxed{2}}$

Simplify each product by making a diagram.

1. $(5x^3)(-2xy^2)$

2. $(a^2b)(2b^4c^3)$

3. $(6p^4q^2)(4pq^3)$

_____   _____   _____

4. $(-3cd^2)(-5c^3)$

5. $(4a^4bc^2)(b^2cd^3)$

6. $(-x^2y^2)(2w^3y)$

_____   _____   _____

7. $(-3mn^3)(-mp^2)$

8. $(5w^2x^2y)(2wx^2y)$

9. $(-3g^3h)(g^2)$

_____   _____   _____

10. $(5p^2q^5r)(-p^3r^2)$

11. $(-hk^2)(-6k^3m)$

12. $(-2b^3c^2d^3)(3a^3bd^2)$

_____   _____   _____

13. Critical Thinking   How can you predict the number of different variables in the product of two monomials?

_____

_____

Problem Solving/Critical Thinking

8.2 *Describing Exponential Expressions*

Here are three exponential properties that you have learned. In each property, x can be any number, and m and n are positive integers.

Product of Powers	Power of a Power	Power of a Product
$x^m \cdot x^n = x^{m+n}$	$(x^m)^n = x^{mn}$	$(xy)^n = x^n y^n$

Before you can evaluate an exponential expression, you must first analyze how it is constructed. Only then can you choose the appropriate property to use.

Describe each exponential expression as a product of powers, a power of a power, or a power of a product.

1. $(p^2)^3$

2. $(p^2)(p^5)$

3. $(-9mp)^2$

_____

_____

_____

4. $(4^2)(4^5)(4^3)$

5. $(4c)^3$

6. $(4^3)^5$

_____

_____

_____

7. $(3xy)^{12}$

8. $(x^3)(x^2)(x^4)$

9. $(x^{10})^2$

_____

_____

_____

To evaluate some exponential expressions, you need to use more than one property. For each expression, describe which properties you would use to evaluate it.

10. $(k^4 k^2)^3$ _____

11. $(p^2)^3 (p^3)^5$ _____

12. $(4^3)^5 (4c)^3$ _____

13. $(6p^4 q^2)^2 (pq^3)$ _____

14. $(6x^2 y)^2 (2x^2 y)^3$ _____

15. Critical Thinking When using more than one property to evaluate an expression, does it matter which property you use first? Justify your answer.

_____

_____

_____

Problem Solving/Critical Thinking

8.3 Using Tables for Monomial Quotients

Here is a way to use a table to organize your work when you are dividing monomials.

◆ **Example**
Simplify this quotient.

$$\frac{-5c^3d}{-15cd^2}$$

◆ **Step 1**
Make the table.

Number	c	d
-5	c^3	d
-15	c	d^2

◆ **Step 2**
Write a fraction for each part of the quotient.

$$\frac{1}{3} \cdot \frac{c^2}{1} \cdot \frac{1}{d} = \frac{c^2}{3d}$$

Complete the table to simplify each quotient. Write your solution below the chart.

1. $\dfrac{-3g^3h}{12g^2}$

Number	g	h

_____

2. $\dfrac{8x^3}{-2xy^2}$

Number	x	y

_____

3. $\dfrac{6p^4q^2}{24pq^3}$

Number	p	q

_____

4. $\dfrac{-20x^2y^2}{4w^3y}$

Number	w	x	y

_____

5. $\dfrac{9p^2q^5r}{-3p^3r^2}$

Number	p	q	r

_____

6. $\dfrac{-3hk^2}{-6k^3m}$

Number	h	k	m

_____

7. $\dfrac{-2b^3c^2d^3}{14b^3cd^2}$

Number	b	c	d

_____

8. Critical Thinking How can you use substitution to check that you have correctly simplified a polynomial quotient?

_____

_____

Problem Solving/Critical Thinking
8.4 Solving a Simpler Problem

Remember that an expression with negative exponents can always be written as an equivalent expression that uses only positive exponents.

◆ **Examples** $-3g^{-3}h = \dfrac{-3h}{g^3}$ $-h^{-1}g^2 = \dfrac{-g^2}{h}$

When you are simplifying complicated expressions that include negative exponents, you may find it helpful to first write the expression as a product of fractions that use only positive exponents.

◆ **Examples** $(-3g^{-3}h)(-h^{-1}g^2) = \dfrac{-3}{1} \cdot \dfrac{1}{g^3} \cdot \dfrac{h}{1} \cdot \dfrac{-1}{h} \cdot \dfrac{g^2}{1} = \dfrac{3g^2h}{g^3h} = \dfrac{3}{g}$

$\dfrac{9p^2q^{-5}r}{-3p^{-3}r^2} = \dfrac{9}{-3} \cdot \dfrac{p^2 \cdot p^3}{1} \cdot \dfrac{1}{q^5} \cdot \dfrac{r}{r^2} = \dfrac{-3p^5}{q^5r}$

Write each expression as a product of fractions with positive exponents. Then simplify.

1. $(p^2q^{-1})(p^{-3}q^4)$ _____

2. $(6x^{-2}y)(2x^{-2}y^{-5})$ _____

3. $(3e^{-4}f^3)(ef^{-2})$ _____

4. $(-5c^{-3})(-15cd^2)$ _____

5. $(-2b^{-3}c^2d^{-3})(14b^{-3}cd^2)$ _____

6. $\dfrac{8x^{-3}}{-2xy^2}$ _____

7. $\dfrac{4a^{-4}bc^2}{8b^2cd^{-3}}$ _____

8. $\dfrac{-20x^2y^{-1}}{4x^{-3}y}$ _____

9. $\dfrac{-2b^{-3}c^2d^{-1}}{14b^{-2}cd^2}$ _____

10. **Critical Thinking** A student simplified the expression in Exercise 9 by first adding the exponents in the numerator and then adding the exponents in the denominator. Explain why this method will not work.

_____

_____

Problem Solving/Critical Thinking

8.5 Simplifying Scientific Notation

When converting very large numbers to scientific notation, you may find it easier to do the work in two steps. First, count the number of zeros and use that number as the exponent.

$$24{,}500{,}000{,}000{,}000{,}000 = 245 \times 10^{14}$$

Next, change the first factor to a number between 1 and 10.

$$24{,}500{,}000{,}000{,}000{,}000 = 245 \times 10^{14} = 2.45 \times 10^{16}$$

When you are converting very small numbers, first write the zeros in groups of 3 or 2.

$$0.00000000101 = 0.000\ 000\ 00\ 101$$
$$= .101 \times 10^{-8}$$
$$= 1.01 \times 10^{-9}$$

Write each number in scientific notation.

1. 7,800,000,000,000,000,000,000

2. 0.0000000000782

_____

_____

3. 46,320,000,000,000

4. 0.00000000000000000000031

_____

_____

5. 7,302,000,000,000,000

6. 0.00000000000000614

_____

_____

For these problems, first write each number in standard form. Then write each in scientific notation.

7. 4.8 billion

8. 32.9 trillion

_____

_____

_____

_____

9. 84 millionths

10. 314 billionths

_____

_____

_____

_____

11. **Critical Thinking** Describe a method that you could use to add the numbers 2.4×10^9 and 1.68×10^8.

_____

_____

Problem Solving/Critical Thinking

8.6 *Analyzing Exponential Data Sets*

To check that a data table describes an exponential function, you can look for the following pattern:
* The *x*-values increase by addition. A constant number is added to each value of *x* to get the next value in the table. The constant can be positive or negative.
* The *y*-values increase by multiplication. Each value of *y* is multiplied by a fixed number to get the next *y*-value.

Therefore, the table describes an exponential function: Each time that 3 is added to *x*, *y* is multiplied by 8.

x	y
0	7
3	56
6	448
9	3584
12	28,672

State whether each table describes an exponential function. Explain why or why not.

1.

x	y
0	3
2	12
4	48
6	192
8	768

_____

_____

2.

x	y
0	234.375
1	93.750
2	37.500
3	15.000
4	6.000

_____

_____

3.

x	y
0	12.000.0
3	1200.0
6	120.0
9	12.0
12	1.2

_____

_____

4.

x	y
10	58
9	71
8	84
7	97
6	110

_____

_____

5. Critical Thinking Find an equation of the form $y = ab^x$ that describes the function shown in Exercise 1.

_____

Problem Solving/Critical Thinking

8.7 *Describing Exponential Number Patterns*

Various types of "special" numbers are exponential numbers. For example, *Fermat numbers* can be described by the function equation $F(n) = 2^{2^n} + 1$, where n is a nonnegative integer.

Find the value of each Fermat number.

1. $F(0)$ _____

2. $F(1)$ _____

3. $F(2)$ _____

4. $F(3)$ _____

5. Make an observation about the first four Fermat numbers.

_____

You, like Fermat, may have noticed that the first four numbers are prime. Fermat, noting that $F(4) = 65,537$ is also prime, went on to conjecture that all numbers of this form are prime.

6. Find the value of $F(5)$. _____

7. Using a calculator, divide your value for $F(5)$ by 641. Was Fermat correct in his conjecture? Justify your answer.

_____

Numbers described by the function $C(n) = n(2^n) + 1$, where n is an integer greater than 1, are called *Cullen numbers*.

Show that each of the following is a Cullen number:

8. 9   **9.** 25   **10.** 65

_____   _____   _____

Find the Cullen number for each given value of *n*.

11. $n = 5$ **12.** $n = 6$ **13.** $n = 7$

_____   _____   _____

14. Critical Thinking How many Cullen numbers do you think there are? Explain your answer.

_____

_____

Problem Solving/Critical Thinking
9.1 Recognizing Like Terms

In order to add and subtract polynomials, you must be able to recognize like terms. Sometimes terms must be simplified before you can recognize like terms.

◆ **Example**
Are $7x^2$ and $(4x)(5x)$ like terms?

◆ **Solution**
Simplify: $(4x)(5x) = 20x^2$
Now you can see that both terms involve the same variable and exponent. They are like terms.

◆ **Example**
Are $(3x)(x)$ and $3x^3$ like terms?

◆ **Solution**
Simplify: $(3x)(x) = 3x^2$
After simplifying, you can see that the exponents are different. So these are not like terms.

Determine whether each set contains like terms or unlike terms.

1. $(4)(6x)$ and $(3x)(5)$ _____

2. $(3a)(3a)$ and $(2)(9a)$ _____

3. $(6x^3)(2x^3)$ and $(x^2)(5x^4)$ _____

4. $6b^2$ and $(3b)(2b)(b)$ _____

5. $(4)(y)(3y^3)$ and $(y)(3y)(5y)$ _____

6. $22r^5$ and $(6r^2)^3$ _____

7. $(d)(d^2)(d^3)$ and $[(d)(2d)]^3$ _____

8. $(8x^7)(x^3)$ and $(3x^7)^3$ _____

9. $(7z^4)^3$ and $(6z^3)^4$ _____

10. $3.8v^4(v^2)$ and $(v^2)(v^2)(v^2)$ _____

Simplify each term and then add or subtract.

11. $(4)(6x) + (3x)(5)$ _____

12. $(2)(9a^2) - (3a)(3a)$ _____

13. $(6x^3 + 8) + (2x)(3x)(4x)$ _____

_____

14. $(4y^4 - 15) - [(3y^2)^2 + 5]$

_____

15. $(8a^2)(2a^2)(2a^2) + 5a(6a)(a^4) + (3a^3)(3a^3)$

_____

16. **Critical Thinking** Examine the equation in the box. If $3^2 + 3^2 + 3^2$ is equal to 3^3, why is $x^2 + x^2 + x^2$ not equal to x^3?

$$3^2 + 3^2 + 3^2 = 3^3$$
$$9 + 9 + 9 = 27$$
$$27 = 27$$

_____

Problem Solving/Critical Thinking

9.2 Recognizing Special Products

When you are multiplying polynomials, there are some special products that you can remember to make the multiplication easier.

Perfect square: $(a + b)(a + b) = (a + b)^2 = a^2 + 2ab + b^2$
Difference of two squares: $(a + b)(a - b) = a^2 - b^2$

◆ **Example**
Determine whether the following product represents a perfect square, a difference of two squares, or neither:

$$(3x + y)(y + 3x)$$

◆ **Solution**
$(y + 3x)$ can be rewritten as $(3x + y)$, so the product can be rewritten as $(3x + y)(3x + y)$, or $(3x + y)^2$. This product is a perfect square.

◆ **Example**
Determine whether the following product represents a perfect square, a difference of two squares, or neither:

$$(4x - 2z)(4x + 2y)$$

◆ **Solution**
The second term in each binomial contains a different variable, so this product is neither a perfect square nor a difference of two squares.

For each product, determine whether it is a perfect square, a difference of two squares, or neither.

1. $(4 + 3x)(4 - 3x)$ _____

2. $(yz + 8x)(8x + yz)$ _____

3. $(6x)(x + 6)$ _____

4. $(2.5y + 7)(2.5y - 7)$ _____

5. $(3b + b^2)(3b + b^2)$ _____

6. $(33 + x)(x - 33)$ _____

7. $(2x - 3)(3x - 2)$ _____

8. $(8z - 5)(8z - 5)$ _____

9. $(2t + 21)(-2t + 21)$ _____

10. $(14x + yz)(14x - xy)$ _____

For each special product, write the formula that you would use to multiply. Then find each product.

11. $(3x + 7)(7 + 3x)$ _____

12. $(y - 12)(y + 12)$ _____

13. $(2x - 5)(5 + 2x)$ _____

14. $(y + 12x)(y + 12x)$ _____

15. **Critical Thinking** In what two other ways could you write the expression $(m + n)^2$?

_____

_____

Problem Solving/Critical Thinking

9.3 Applying the FOIL Method

When you are multiplying polynomials, it is often helpful to draw arrows between terms that will be multiplied together when using the FOIL method.

The FOIL method: Multiply the **First** terms.
Multiply the **Outside** terms.
Multiply the **Inside** terms. Add the outside and the inside products.
Multiply the **Last** terms.

◆ **Example**
Draw arrows to show each step of the FOIL method. Then find the product.
$$(3x + 5y)(x + 3y)$$

◆ **Solution**

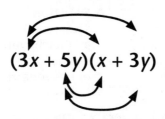

$(3x + 5y)(x + 3y)$ Draw arrows for each product.

$(3x)(x) + (3x)(3y) + (x)(5y) + (5y)(3y)$ Write out each product.

$3x^2 \quad + 9xy \quad + \quad 5xy \quad + \quad 15y^2$ Simplify each product.

$3x^2 \quad + 14xy \qquad\qquad\quad + \quad 15y^2$ Add the outside and the inside products.

Draw arrows to show each step of the FOIL method. Then find each product.

1. $(3x + 1)(x + 3)$ _____

2. $(y - 12)(2y - 9)$ _____

3. $(2x - 5)(x + 8)$ _____

4. $(y + 4x)(y + 4x)$ _____

5. $(2b + 5c)(4b + 3c)$ _____

6. Critical Thinking The expression at $(2x)(3x) + (2x)5 + 7(3x) + 7(5)$
right is the result of applying the FOIL
method. Use arrows to find the original
product. Then write the product of
the two binomials that creates this expression. _____

NAME _____ CLASS _____ DATE _____

Problem Solving/Critical Thinking
9.4 Using Visual Verification

When you are working with polynomials, you can use visual aids to verify
that two polynomial expressions are equal. By graphing both expressions,
you can tell whether they share the same solutions. If the graphs are the
same, then the expressions are equal.

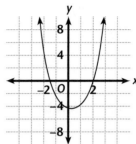

◆ **Example**
Use visual verification to determine if $(2x + 3)(x - 2)$
equals $2x^2 - x - 6$.

◆ **Solution**
Enter both expressions into a graphics calculator.
Compare the graphs. If you see only one graph,
then the two graphs lie on top of each other and,
therefore, are equal, so the equation is true.

**Use visual verification to determine whether each equation is
true. Set each side of the equation equal to y and then sketch the
graph(s) on each grid. Write true or false next to each equation.**

1. $(4x - 3)(x - 3) = 4x^2 + 6x - 9$ _____

2. $(x + 5)(x - 5) = x^2 - 25$ _____

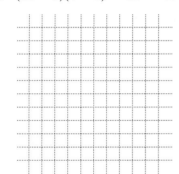
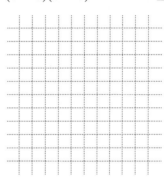

3. $(x + 3)(x + 3) = x^2 + 6x + 9$ _____

4. $(x + 2)(2x - 3) = 2x^2 + x - 5$ _____

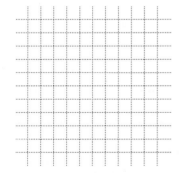
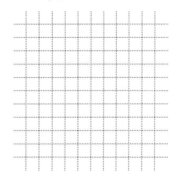

5. Critical Thinking When a student entered the two
expressions at right into a graphics calculator, she
only saw one graph. The expressions, however, are not
equal. How can you explain this?

$(x + 1)(x + 1)$ and $x^2 + 2x + 14$

_____

Algebra 1   **Problem Solving/Critical Thinking**   **53**

Problem Solving/Critical Thinking
9.5 Solving a Simpler Problem: Factoring

Sometimes when you are factoring polynomials, the GCF is not easy to find.
In these cases, you can factor the polynomial one expression at a time.

◆ **Example**
Factor the polynomial $4x^3 + 16x^2$

◆ **Solution**
Factor out the 4: $4(x^3 + 4x^2)$
Then factor out the x^2: $4x^2(x + 4)$

◆ **Example**
Factor the polynomial $75a^3 + 30a^2 + 15a$

◆ **Solution**
Factor out the 15: $15(5a^3 + 2a^2 + a)$
Then factor out the a: $15a(5a^2 + 2a + 1)$

Factor each polynomial one factor at a time.

1. $4ab^2 + 12a$ _____

2. $25x^2 - 15x$ _____

3. $17c^4 + 51c^2$ _____

4. $9w^3 + 12w^2 + 21w$ _____

5. $10y^6 + 8y^4 - 6y^2$ _____

6. $5mn^2 - 40m^2$ _____

7. $6xy^2z + 15xy$ _____

8. $19ab^2c^3 - 35a^3b^2c$ _____

9. $4c^5d^3 + 12c^2d$ _____

10. $22p^2r - 55pqr$ _____

11. **Critical Thinking** What is the significance of the expression that
you factored out of the polynomial?

_____

_____

_____

Problem Solving/Critical Thinking

9.6 *Recognizing Terms in Special Polynomials*

Factoring special polynomials requires the identification of special terms. You must be able to recognize terms that are perfect squares. In perfect-square trinomials, you must also be able to recognize the special middle term. This term is twice the product of the roots of the perfect-square terms.

◆ **Example**
If the polynomial looks like a perfect-square trinomial, find the square roots of the first and the last terms. Verify that the middle term is twice their product.

$9p^2 - 30pq + 25q^2$

◆ **Solution**
First root: $3p$
Second root: $5q$
Twice the product: $(2)(3p)(5q) = 30pq$
So, this polynomial represents a perfect-square trinomial.

◆ **Example**
If the polynomial looks like a difference of two squares, find the square roots of both terms.

$4x^4 - 9y^2$

◆ **Solution**
First root: $2x^2$
Second root: $3y$
So, this polynomial is a difference of two squares.

For each pair of terms, circle the term that is the perfect square.

1. $12x^2$ or $16y^2$

2. $81b^4$ or $111d^2$

3. $36a$ or $9g^2$

4. $25m^6$ or $44n^2$

5. r^3 or 100

6. x^4 or ab^2

For each special polynomial, find the roots of the first and the last term. If there is a middle term, verify that it is twice the product of the roots.

7. $4x^2 - 16$

roots: _____ _____

8. $25c^4 - 144x^2$

roots: _____ _____

9. $36a^2 + 12ac + c^2$

roots: _____ _____

twice the product: _____

10. $64x^2 - 176xy + 121y^2$

roots: _____ _____

twice the product: _____

11. Critical Thinking Why is the polynomial $25x^2 + 20xy + 16y^2$ not a perfect-square trinomial?

_____

_____

_____

Problem Solving/Critical Thinking

9.7 Trial and Error in Factoring Trinomials

Factoring trinomials requires you to find possible factor combinations and to use trial and error to discover which combination is correct.

◆ **Example**
Factor the following trinomial:
$$x^2 - 6x + 8$$

◆ **Solution**
Possible factor combinations are 8 and 1, −8 and −1, 4 and 2, −4 and −2.
Which combination adds up to the constant of the middle term, −6?
The correct combination is −4 and −2.
So the trinomial factors are $(x - 4)$ and $(x - 2)$.

◆ **Example**
Factor the following trinomial:
$$a^2 + 5a - 6$$

◆ **Solution**
Possible factor combinations are 6 and −1, −6 and 1, 3 and −2, −3 and 2.
Which combination adds up to the constant of the middle term, 5?
The correct combination is 6 and −1.
The trinomial factors are $(a + 6)$ and $(a - 1)$.

List all of the possible factor combinations for each number.

1. 18 _____

2. 24 _____

3. −12 _____

4. −30 _____

For each polynomial, list the possible factor combinations of the last term. Then circle the combination whose sum equals the constant of the middle term.

5. $x^2 + 8x + 15$ _____

6. $y^2 - 4y - 12$ _____

7. $r^2 + 21r - 100$ _____

8. $a^2 - 19a + 48$ _____

9. $c^2 - c - 42$ _____

10. **Critical Thinking** For the polynomial $x^2 + 19x + 84$, how can you immediately eliminate all negative factor combinations?

_____

_____

Problem Solving/Critical Thinking

9.8 Using Lines to Solve Quadratic Equations

When you use factoring to solve a quadratic equation, you replace a single quadratic equation with a pair of linear equations, which are easier to solve than the original.

◆ **Example**
Solve the following equation by factoring:
$$x^2 - 5x + 6 = 0$$

◆ **Solution**
Factor the trinomial: $(x - 3)(x - 2)$
Turn each factor into a
linear equation equal to 0.
$$x - 3 = 0 \text{ and } x - 2 = 0$$

Solve for x in each equation: 3 and 2

◆ **Example**
Solve the following equation by factoring:
$$a^2 + 10a + 24 = 0$$

◆ **Solution**
Factor the trinomial: $(a + 6)(a + 4)$
Turn each factor into a
linear equation equal to 0.
$$a + 6 = 0 \text{ and } a + 4 = 0$$

Solve for a in each equation: -6 and -4

Write each of the factored quadratic equations as a pair of linear equations that you could solve to find the solution(s) to the given equation.

1. $(x + 7)(x - 9) = 0$ _____

2. $(n - 3)(n - 8) = 0$ _____

3. $(5a - 2)(2a + 9) = 0$ _____

4. $(3y + 4)(4y + 5) = 0$ _____

Solve each quadratic equation by factoring. Use the factors to create two linear equations. Solve those equations.

5. $x^2 + 12x + 20 = 0$ _____

6. $y^2 - 2y - 15 = 0$ _____

7. $r^2 - 17r + 30 = 0$ _____

8. $a^2 + 8a - 48 = 0$ _____

9. $d^2 - d - 90 = 0$ _____

10. **Critical Thinking** What does the factored quadratic equation
$y = (2x + 3)(x + 2)$ have in common with the lines whose equations
are $y = 2x + 3$ and $y = x + 2$?

_____

_____

Problem Solving/Critical Thinking
10.1 *Visualizing Transformations*

When you are trying to visualize a quadratic equation, you can often begin with the parent function and then perform transformations one at a time until you get the graph of the original equation. The movement of a special point using arrows will help.

◆ **Example**
Graph $y = -(x - 3)^2 + 2$ using transformations one at a time. Use the vertex and arrows as a guide.

◆ **Solution**

a. Roughly represent the graph of $y = x^2$. Mark the vertex and sketch an upward U.

b. Sketch an upward arrow 2 units long and move the U 2 units up.

c. Sketch a right arrow 3 units long and move the U 3 units to the right.

d. Flip the U over.

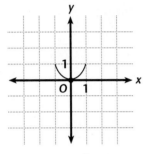

$y = x^2$

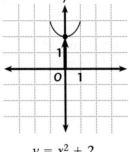

$y = x^2 + 2$

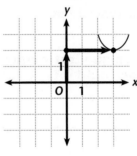

$y = (x - 3)^2 + 2$

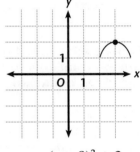

$y = -(x - 3)^2 + 2$

The graph is a parabola opening downward with a vertex of $(3, -2)$.

Graph each equation by performing transformations one at a time. Describe each transformation; write the correct equation, and sketch the graph. Use the vertex and arrows as guides.

1. $y = -(x + 2)^2 + 1$

a. transformation:

_____

equation:

_____

b. transformation:

_____

equation:

_____

c. transformation:

_____

equation:

_____

d. transformation:

_____

equation:

_____

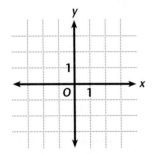

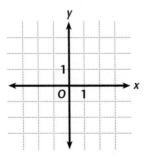

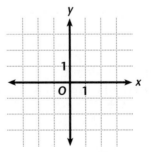

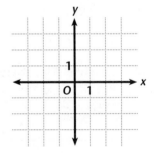

2. Critical Thinking Can you perform the transformations in any order? Explain.

_____

Problem Solving/Critical Thinking

10.2 *Recognizing Patterns in Solutions*

All equations of the form $ax^2 = k$ can be solved in the same way.

$$x = \sqrt{\frac{k}{a}} \text{ or } x = -\sqrt{\frac{k}{a}}$$

◆ **Example**
Solve the equation $4x^2 = 49$.

◆ **Solution**
Isolate the variable.

$$x^2 = \frac{49}{4}$$

Take the square root of each side.　$x = \sqrt{\frac{49}{4}}$ or $x = -\sqrt{\frac{49}{4}}$

Simplify.　　　　　　　　　　　$\frac{7}{2}$ and $-\frac{7}{2}$

Using the example above as a guide, solve each equation. Write your work in the space provided.

1.　$25x^2 = 16$ _____

2.　$9x^2 = 64$ _____

3.　$4x^2 = 100$ _____

4.　$16x^2 = 121$ _____

5.　$81x^2 = 1$ _____

6.　$64x^2 = 144$ _____

7.　Establish a pattern in the operations that you used to solve each equation in Exercises 1–6. Write out the steps for solving an equation in
the form $ax^2 = k$, where $\frac{k}{a} > 0$ and $a \neq 0$.

_____

_____

8.　**Critical Thinking** Why can equations of the form $ax^2 = k$ only be
solved if $\frac{k}{a} > 0$ and $a \neq 0$.

_____

_____

Problem Solving/Critical Thinking

10.3 Using Algebra to Predict Graphs

The process of adding neutral pairs of numbers to an equation is an algebraic technique that can be used to identify some characteristics of a graph. For quadratic equations, you can use the technique of completing the square to find the vertex. When you know the vertex, you can write an equation for the axis of symmetry.

◆ **Example**

Find the coordinates of the vertex and an equation for the the axis of symmetry of the parabola described by the equation $y = x^2 + 6x - 5$.

◆ **Solution**

Complete the square.

$y = (x^2 + 6x) - 5$	Group the terms with variables.
$\left(\frac{6}{2}\right)^2 = (3)^2 = 9$	Find the number that will complete the square.
$y = (x^2 + 6x + 9) - 5 - 9$	Add neutral pairs containing the number.
$y = (x + 3)^2 - 14$	Rewrite in the form $y = (x - h)^2 + k$.

The vertex is $(-3, -14)$. The axis of symmetry is the line $x = -3$.

Use the method of completing the square to rewrite each function in the form $y = (x - h)^2 + k$. Then find the coordinates of the vertex of each parabola and write an equation for the axis of symmetry.

1. $y = x^2 - 8x + 10$ _____

   vertex: _____

   axis of symmetry: _____

2. $y = x^2 + 12x + 27$ _____

   vertex: _____

   axis of symmetry: _____

3. $y = x^2 - 5x + 6$ _____

   vertex: _____

   axis of symmetry: _____

4. $y = x^2 + 20x + 99$ _____

   vertex: _____

   axis of symmetry: _____

5. **Critical Thinking** The equation $y = x^2 + bx + c$ is written in the form $y = (x - h)^2 + k$. How are h and b related? k and c related?

   _____

   _____

   _____

Problem Solving/Critical Thinking

10.4 *Choosing the Best Method of Solution*

So far, you have learned to solve quadratic equations by factoring, by completing the square, and by graphing. By closely inspecting an equation, you can choose the quickest and easiest method of solution.

◆ **Example**
Solve the equation $x^2 + 3x - 10 = 0$.

◆ **Solution**
This equation easily factors, but attempting to complete the square would require a fraction. The best method here is factoring.

$$x^2 + 3x - 10 = 0$$
$$(x + 5)(x - 2) = 0$$
$$x = -5 \text{ or } x = 2$$

The solutions are −5 and 2.

◆ **Example**
Solve the equation $x^2 + 6x - \frac{13}{4} = 0$.

◆ **Solution**
The fraction in the equation makes factoring difficult, but the coefficient of the x-term makes completing the square relatively easy.

$$(x^2 + 6x + \mathbf{9}) - \frac{13}{4} - \mathbf{9} = 0$$
$$(x + 3)^2 = \frac{49}{4}$$
$$x + 3 = \pm\frac{7}{2}$$
$$x = \frac{1}{2} \text{ and } x = -\frac{13}{2}$$

◆ **Example**
Solve the equation $x^2 + 1.5x - 0.60 = 0$.

◆ **Solution**
The decimals in the equation make graphing the best method of solution in this case.

For each equation, choose the best method of solution.

1. $x^2 - 8x + 15 = 0$ _____

2. $x^2 - 1.2x - 0.36 = 0$ _____

3. $x^2 - 3x - \frac{27}{4} = 0$ _____

4. $3x^2 - 4x - 6 = 0$ _____

5. $x^2 + \frac{3}{2}x - 1 = 0$ _____

6. $x^2 - 11x + 30 = 0$ _____

Choose any method to solve each equation.

7. $x^2 + 6x - 7 = 0$ _____

8. $x^2 + \frac{2}{3}x - \frac{8}{9} = 0$ _____

9. $x^2 + 4x + 6 = 0$ _____

10. $x^2 - 10x + 24 = 0$ _____

11. **Critical Thinking** Complete the square to show that the solutions to $x^2 + bx + c = 0$ are $-\frac{b}{2} + \sqrt{\frac{b^2 - 4c}{4}}$ and $-\frac{b}{2} - \sqrt{\frac{b^2 - 4c}{4}}$.

_____

_____

Problem Solving/Critical Thinking

10.5 *Using the Quadratic Formula to Solve Related Equations*

Using the quadratic formula and a graphics calculator, you can solve a set
of related equations quickly and easily. Once you have entered the formula
into the calculator, you can reuse the formula with a little editing.

◆ **Example**
   Use the quadratic formula to solve each equation.

    a. $x^2 + 5x + 2 = 0$ **b.** $x^2 + 5x + 3 = 0$ **c.** $x^2 + 5x + 4 = 0$

◆ **Solution**

    a. Enter the formula into a graphics calculator with $a = 1$, $b = 5$, and $c = 2$. Reuse the formula by replacing the plus sign with a minus sign.

    b. The only change in the equation is the value of c. Reuse the formula to find both solutions by making the appropriate edits.

    c. Repeat the process with the new value $c = 4$.

```
(-5+r(5²-4*1*2))
/(2*1)
         -.4384471872
(-5-r(5²-4*1*2))
/(2*1)
         -4.561552813
```

```
(-5+r(5²-4*1*3))
/(2*1)
         -.6972243623
(-5-r(5²-4*1*3))
/(2*1)
         -4.302775638
■
```

```
(-5+r(5²-4*1*4))
/(2*1)
                  -1
(-5-r(5²-4*1*3))
/(2*1)
                  -4
```

If the calculator gives you an error, first make sure you have entered the
formula correctly. If you still receive an error, then there is no solution to
the equation.

**Use a graphics calculator and the quadratic formula to find all the solutions to each
equation.**

1.   **a.** $4x^2 + 6x - 9 = 0$ _____

    b. $4x^2 + 6x - 7 = 0$ _____

    c. $4x^2 + 6x - 5 = 0$ _____

2.   **a.** $2x^2 + 5x + 2 = 0$ _____

    b. $2x^2 + 5x + 3 = 0$ _____

    c. $2x^2 + 5x + 4 = 0$ _____

3.   **Critical Thinking** The equation $ax^2 + (a + 1)x + (a + 2) = 0$ has
exactly one solution for x. Find a. Show your work below.

_____

_____

Problem Solving/Critical Thinking

10.6 *Using Test Points to Verify Solutions*

Given a quadratic inequality in x and y, an ordered pair either satisfies the inequality or does not.

♦ If it does, the solution is the region which contains the point that corresponds to the ordered pair.

♦ If it does not, the solution region is the part of the plane that does not contain the point that corresponds to the ordered pair.

♦ **Example**
Graph the solution to $y > 2x^2 - 5x + 3$ by using a test point.

♦ **Solution**
Graph the solution.

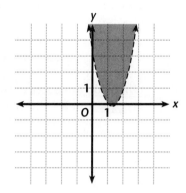

Choose a point such as $(1, 2)$. Substitute the coordinates into the inequality and simplify.

$$2 > 2(1)^2 - 5(1) + 3$$
$$2 > 0 \qquad \text{true}$$

The solution is the part of the plane determined by the graph of $y = 2x^2 - 5x + 3$ that contains the point $(1, 2)$.

Graph each inequality by using a test point.

1. $y < -x^2 + 3x - 3$

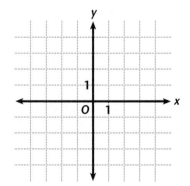

test point: _____

solution: _____

2. $y < 4x^2 - 9x + 2$

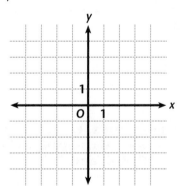

test point: _____

solution: _____

3. $y \geq -3x^2 + 7x + 4$

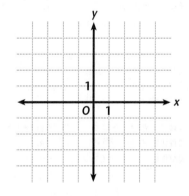

test point: _____

solution: _____

4. Critical Thinking Refer to the statements at right. What can you say about statement B? How does this relate to graphing quadratic inequalities?

> Either statement A is true or statement B is true. Both statements cannot be true simultaneously. Statement A is false.

_____

_____

Problem Solving/Critical Thinking

11.1 *Evaluating Inverse Variation Relationships*

When you are working with inverse variation, you can use one ordered pair of x and y to write an equation that expresses the relationship. You can then evaluate the equation for other values of x.

◆ **Example**
   If y varies inversely with x, and $y = 15$ when $x = 4$, find the value of y
   when $x = 1, 2, 3, 4,$ and 5.

◆ **Solution**
   Because y varies inversely with x, we know that the constant of
   variation, k, is equal to the product of x and y.

   $$k = xy$$
   $$k = (4)(15)$$
   $$k = 60$$

   Once you have found the constant of variation, you can write
   an equation for this inverse variation relationship.
   Then evaluate the expression for each x-value.

   $$y = \frac{60}{x}$$

For $x = 1$:	For $x = 2$:	For $x = 3$:	For $x = 4$:	For $x = 5$:
$y = \dfrac{60}{1}$	$y = \dfrac{60}{2}$	$y = \dfrac{60}{3}$	$y = \dfrac{60}{4}$	$y = \dfrac{60}{5}$
$y = 60$	$y = 30$	$y = 20$	$y = 15$	$y = 12$

The solution can be expressed using the ordered pairs $(1, 60)$, $(2, 30)$,
$(3, 20)$, $(4, 15)$, $(5, 12)$.

**In Exercises 1– 4, y varies inversely with x. Find the constant of
variation, k, for each set of values.**

1. $x = 6$ when $y = 5$ _____   **2.** $x = 2$ when $y = 16$ _____

3. $x = 8$ when $y = 35$ _____   **4.** $x = 12$ when $y = 20$ _____

**In Exercises 5– 7, y varies inversely with x. Write an equation
that describes each inverse variation relationship. Then
evaluate each equation for $x = 1, 2, 3, 4,$ and 5.**

5. $x = 6$ when $y = 6$ _____

6. $x = 12$ when $y = 20$ _____

7. $x = -9$ when $y = 10$ _____

8. Critical Thinking If y varies inversely with x, what happens to the
   value of y as the value of x approaches 1? What happens to the value
   of y as the value of x approaches zero?

_____

Problem Solving/Critical Thinking

11.2 *Choosing the Best Method of Solution*

When you are working with rational functions it is important to determine whether the function is undefined for any values of x. Because division by zero is undefined, a rational function is undefined when its denominator is equal to zero.

In some cases you can tell by inspection which x-values will make the denominator zero. In other cases you will need to use simple algebra to determine these values. In still other cases you may have to factor or use the quadratic formula to find these values.

◆ **Example**

Find any x-values for which each rational function is undefined.

a. $f(x) = \dfrac{3}{x^2 + 1}$
    b. $g(x) = \dfrac{7x + 5}{2x - 1}$
    c. $h(x) = \dfrac{3x + 5}{x^2 - 7x + 6}$

◆ **Solution**

a. By inspection, we can see that the denominator will always be positive and will never be zero. Therefore, this function is defined for all values of x.

b. Here you can set the denominator equal to zero and solve $2x - 1 = 0$ mentally. This function is undefined at $x = \dfrac{1}{2}$.

c. Here you must solve the quadratic expression in the denominator. Since this denominator easily factors to $(x - 1)(x - 6)$, you find that the function is undefined when $x = 1$ or when $x = 6$.

For each rational function, find any values for x where the function is undefined. Write the method that you used to determine the solution.

1. $d(x) = \dfrac{3x + 2}{x + 5}$ _____

2. $m(x) = \dfrac{15x - 1}{2x^2 + 3}$ _____

3. $p(x) = \dfrac{18x}{x^2 - 4}$ _____

4. $f(x) = \dfrac{x^2 - 10}{x^2 + 3x - 28}$ _____

5. Critical Thinking What conclusions can you draw about the domain of a rational function with a denominator in the form $x^2 + a$ when a is positive?

_____

Problem Solving/Critical Thinking

11.3 Exploring Shortcuts to Simplifying Rational Expressions

Simplifying rational expressions can be a time-consuming and difficult process. Using logic and your knowledge of factors, you can make the process faster and easier. In some cases, you can use inspection to cancel out a common factor. In other cases, you can use some simple factoring skills to find that a rational expression cannot be simplified.

◆ **Example**

Simplify the rational expression if possible:

$$\frac{3x^2 - 5x}{x^3 + 2x}$$

◆ **Solution**

A quick inspection of the expression shows that every term contains an x, so this is a common factor. Canceling one x from each term gives the simplified expression $\frac{3x - 5}{x^2 + 2}$.

◆ **Example**

Simplify the rational expression if possible:

$$\frac{x^2 - 2x - 15}{x^2 + 6x + 8}$$

◆ **Solution**

Examine the factor pairs for the last term in each expression. The factor pairs for -15 are $(1, -15), (-1, 15), (3, -5)$, and $(-3, 5)$. The factor pairs of 8 are $(1, 8), (-1, -8), (2, 4)$, and $(-2, -4)$. The only possible common factors are 1 and -1. Using either of these factors in the numerator will not give the correct coefficient of the middle term, -2. This expression cannot be simplified.

Use shortcuts to simplify each rational expression if possible. If the expression cannot be simplified, explain why not.

1. $\dfrac{2b + 6}{4b^3 + 10}$ _____

2. $\dfrac{t^2 - 9}{t^2 - 3t - 10}$ _____

3. $\dfrac{8x^3 + 3x^2 + 7x}{13x^4 - 6x^2}$ _____

4. $\dfrac{m^2 + 6m + 8}{m^2 - 4m - 21}$ _____

5. **Critical Thinking** Can the rational expression $\dfrac{b^2 + 7b + 6}{c^2 + 3c + 2}$ be simplified? Explain.

_____

_____

_____

Problem Solving/Critical Thinking

11.4 *Using Tables to Add Rational Expressions*

When you are performing operations on rational expressions, you will use many of the same steps you use when you are working with fractions. To add and subtract rational expressions, you must find a common denominator, change both expressions to common denominators, and then simplify.

You can use a table to demonstrate how the process works for fractions and rational expressions. Substitute a simple value for the variable in each expression in order to form a simple fraction. You can perform the same operations on the fraction and the rational expression to find the solution.

◆ **Example**
  Use a table to find $\dfrac{(x-3)}{3x} + \dfrac{(x+1)}{5x}$. Show the work for simple fractions and for the rational expressions.

◆ **Solution**
  First, form simple fractions by substituting 10 for x. $\dfrac{(10-3)}{30} + \dfrac{(10+1)}{50} = \dfrac{7}{30} + \dfrac{11}{50}$

  Then, create a table to perform the necessary steps.

Fractions	Rational Expressions	
$\dfrac{7}{30} + \dfrac{11}{50}$	$\dfrac{(x-3)}{3x} + \dfrac{(x+1)}{5x}$	Given
$\dfrac{5}{5} \cdot \dfrac{7}{30} + \dfrac{3}{3} \cdot \dfrac{11}{50}$	$\dfrac{5}{5} \cdot \dfrac{(x-3)}{3x} + \dfrac{3}{3} \cdot \dfrac{(x+1)}{5x}$	Find a common denominator.
$\dfrac{35}{150} + \dfrac{33}{150}$	$\dfrac{5x-15}{15x} + \dfrac{3x+3}{15x}$	Change to common denominators.
$\dfrac{68}{150} = \dfrac{34}{75}$	$\dfrac{8x-12}{15x}$	Simplify.

Finally, substitute the same value into the simplified

rational expression to check your solution. $\dfrac{8(10)-12}{150} = \dfrac{68}{150} = \dfrac{34}{75}$

Use a table to add each pair of rational expressions. Show the work for simple fractions and for the rational expressions.

1. $\dfrac{x+2}{2x} + \dfrac{x-4}{3x}$ _____

2. $\dfrac{x+5}{3x} + \dfrac{x-2}{9x}$ _____

3. $\dfrac{x+1}{5x} + \dfrac{3x-2}{4x}$ _____

Problem Solving/Critical Thinking

11.5 *Using a Proportion to Solve a Rational Equation*

Some rational equations can be solved by using proportions. You can simplify both sides of the equation to set up the proportion. Once the proportion is set up, you can use cross multiplication to eliminate the denominators. The resulting equation can be solved by using linear or quadratic equations.

◆ **Example**
   Use a proportion to solve the following rational equation:

$$\frac{x-2}{2x} + \frac{x+4}{3x} = \frac{2x}{x+2}$$

◆ **Solution**
   First, simplify the left side of the equation:

$$\frac{3}{3} \cdot \frac{x-2}{2x} + \frac{2}{2} \cdot \frac{x+4}{3x} = \frac{2x}{x+2}$$

$$\frac{3x - 6 + 2x + 8}{6x} = \frac{2x}{x+2}$$

$$\frac{5x+2}{6x} = \frac{2x}{x+2}$$

Then cross multiply: $(x+2)(5x+2) = (6x)(2x)$

Simplify: $5x^2 + 12x + 4 = 12x^2$

$7x^2 - 12x - 4 = 0$

Solve the resulting quadratic equation: $(7x+2)(x-2) = 0$

$$x = -\frac{2}{7} \text{ or } x = 2$$

Use a proportion to solve each rational equation.

1. $\dfrac{x+3}{x} = \dfrac{x+11}{3x-5}$ _____

2. $\dfrac{x+1}{x} + \dfrac{x-2}{5x} = \dfrac{2x+1}{x+2}$ _____

3. $\dfrac{x-3}{4x} + \dfrac{x+1}{2x} = \dfrac{2x+6}{x}$ _____

4. $\dfrac{x-2}{3x} + \dfrac{x+4}{5x} = \dfrac{x-1}{x+3}$ _____

5. **Critical Thinking** How can you tell by inspection that Exercise 3 only has one solution?

_____

_____

Problem Solving/Critical Thinking

11.6 Related Conditional Statements

A statement in if-then form is called a *conditional statement.*   If *p,* then *q.*

The part of the statement that follows *if* is the hypothesis, and the part that follows *then* is the conclusion. When the hypothesis and conclusion of the original statement are interchanged, the new statement is the *converse* of the original statement.   If *q,* then *p.*

When the hypothesis and conclusion of the original statement are negated, the new statement is the *inverse* of the original statement.   If not *p,* then not *q.*

When the hypothesis and conclusion of the converse are negated, the new statement is the *contrapositive* of the original statement.   If not *q,* then not *p.*

Consider the following conditional statement: If a polygon has three sides, then it is a triangle.

1. The converse is _____.

2. The inverse is _____.

3. The contrapositive is _____.

4. The original conditional is a true statement. Which of the related
   conditional statements are also true?   _____

Consider the following statement: If a polygon is a rectangle, then it is a quadrilateral.

5. The converse is _____.

6. The inverse is _____.

7. The contrapositive is _____.

8. The original conditional is a true statement. Which of the related
   conditional statements are also true?   _____

Consider the following statement: If two angles are right angles, then they are equal in measure.

9. The converse is _____.

10. The inverse is _____.

11. The contrapositive is _____.

12. The original conditional is a true statement. Which of the related
    conditional statements are also true?   _____

Problem Solving/Critical Thinking

12.1 Rearranging a Problem

When you are working with square roots, it often helps to rearrange the problem so that you can easily gather like terms. An easy way to do this is to rewrite problems vertically.

◆ **Example**
Simplify the radical expression. $8\sqrt{3} + 5\sqrt{2} - 3\sqrt{3} + 2\sqrt{2}$

◆ **Solution**
Rewrite the problem vertically.

$$8\sqrt{3} + 5\sqrt{2}$$
$$\underline{-3\sqrt{3} + 2\sqrt{2}}$$

Then add or subtract like terms. $5\sqrt{3} + 7\sqrt{2}$

Rearrange each radical expression in vertical form.

1. $7\sqrt{5} + 3\sqrt{2} + 9\sqrt{5} + 2\sqrt{2}$

2. $3\sqrt{7} + 5\sqrt{3} - (2\sqrt{7} + 4\sqrt{3})$

3. $14\sqrt{3} - 5\sqrt{2} + 8\sqrt{3} + 10\sqrt{2}$

4. $9\sqrt{2} + 3\sqrt{2} - 3\sqrt{5} + 6\sqrt{2}$

5. $17\sqrt{2} + 21\sqrt{3} + 33\sqrt{5} + 3\sqrt{2} - 6\sqrt{3} - 3\sqrt{5}$

Simplify each rearranged expression.

6. Exercise 1 _____

7. Exercise 2 _____

8. Exercise 3 _____

9. Exercise 4 _____

10. Exercise 5 _____

11. **Critical Thinking** The radical expression $\sqrt{a} + \sqrt{b}$ cannot always be rewritten as $\sqrt{a + b}$. Find three ordered pairs (a, b) such that $\sqrt{a} + \sqrt{b} = \sqrt{a + b}$. Find all a and b such that the equation is true.

_____

_____

_____

Problem Solving/Critical Thinking

12.2 *Using Quadratics to Solve Radical Equations*

When you are solving radical equations, you can square both sides of the equation to make it easier to solve. Sometimes, the resulting equation will be a quadratic equation. You can apply the skills that you have already learned to solve these quadratic equations.

When you use this method to solve radical equations, you must check your answer(s) to make sure that each answer is a true solution.

◆ **Example**

Solve $\sqrt{x + 6} = x$.

◆ **Solution**

Square both sides:	$(\sqrt{x + 6})^2 = x^2$
Simplify:	$x + 6 = x^2$
Rearrange:	$0 = x^2 - x - 6$
Factor:	$0 = (x - 3)(x + 2)$
Solve:	$x - 3 = 0$ or $x + 2 = 0$
	$x = 3$ $x = -2$

Check both solutions:

$$\sqrt{3 + 6} = 3 \qquad \sqrt{(-2) + 6} = -2$$
$$\sqrt{9} = 3 \qquad \sqrt{4} = -2$$
$$3 = 3 \qquad 2 \neq -2$$

The solution is $x = 3$. The value of x cannot be -2 because $\sqrt{4}$ is defined as the positive square root of 4.

Solve each radical equation using the method described above.

1. $\sqrt{x + 2} = x$ _____

2. $\sqrt{x + 30} = x$ _____

3. $\sqrt{8x - 15} = x$ _____

4. $\sqrt{5x + 24} = x$ _____

5. $\sqrt{6x - 8} = x$ _____

6. **Critical Thinking** All of the equations above are of the form $\sqrt{ax + b} = x$. Write a generalization by which you can solve all equations of this form.

_____

_____

Problem Solving/Critical Thinking
12.3 Using Formulas to Solve Right-Triangle Problems

Many problems involving right triangles can be solved by applying the Pythagorean Theorem. By solving the Pythagorean Theorem for one variable, you can create a formula that will solve all the problems of the same type.

Suppose that in a right triangle, the legs have lengths of a and of b and the hypotenuse has a length of c. To solve for an unknown length of a leg, a, in a right triangle, you can solve the Pythagorean Theorem for the variable a.

$$c^2 = a^2 + b^2$$
$$c^2 - b^2 = a^2$$
$$\sqrt{c^2 - b^2} = \sqrt{a^2}$$
$$\sqrt{c^2 - b^2} = a$$

This formula can be used to solve all right triangle problems involving an unknown leg.

Label each side of each right triangle as a "leg" or a "hypotenuse."

1.    **2.** 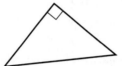   **3.**

4. Explain the meaning of the formula for a found above. _____

Use the formula above to solve for the unknown leg in each triangle.

5. 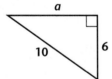   **6.** 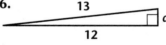   **7.**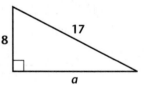

_____   _____   _____

8. Critical Thinking For each triangle below, explain why the formula for a derived above would not solve for the unknown side.

a. 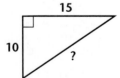   **b.**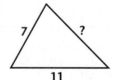

_____   _____

Problem Solving/Critical Thinking

12.4 *Finding Distances: Choosing the Best Method of Solution*

When you are working real world problems involving distance, you can use a coordinate grid to help you visualize the problem. Once you have visualized the problem, the answer may be apparent without performing any calculations. If the answer is not apparent, you can apply the distance formula to calculate any distances that you need to know.

◆ **Example**
A camper wants to hike to the closest outpost. Outpost 14 is 6 miles north and 2 miles east. Outpost 29 is 4 miles north and 4 miles east. If the camper can hike directly to each outpost, which outpost should the camper choose?

◆ **Solution**
First, plot the points on a coordinate grid with the camper's current location as the origin.

Outpost 14 will be represented by the point (2, 6).
Outpost 29 will be represented by the point (4, 4).

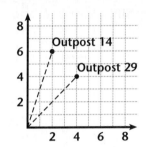

Because the distances are not easy to compare visually, we will apply the distance formula to both possible routes.

Outpost 14: $d = \sqrt{(2 - 0)^2 + (6 - 0)^2}$ Outpost 29: $d = \sqrt{(4 - 0)^2 + (4 - 0)^2}$

$d = \sqrt{4 + 36}$ $d = \sqrt{16 + 16}$

$d = \sqrt{40}$ $d = \sqrt{32}$

Now it is apparent that Outpost 29 is closer and would be the better choice.

Use the grid at right to complete each exercise. For each answer, write whether you found the answer visually or by using the distance formula.

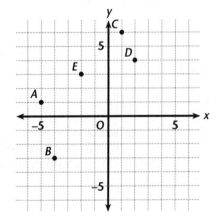

1. Which point is closer to the origin, *A* or *B*?

   _____

2. Which point is closer to the origin, *C* or *D*?

   _____

3. Which point is closest to point *E*?

   _____

4. **Critical Thinking** Explain how to use a compass to solve Exercise 1.

   _____

   _____

Problem Solving/Critical Thinking
12.5 Using Algebra to Solve Geometric Problems

Many geometric problems can be solved by using algebraic formulas that have already been derived. Learning to recognize and apply formulas is an excellent problem-solving skill.

Once you have written the correct formula, you can use the formula to determine what values of x and y are solutions to the problem.

◆ **Example**
Does the point $(7,4)$ lie on the circle that has a radius of 5 and a center $C(3,1)$?

◆ **Solution**

Write the general formula for a circle.	$(x - h)^2 + (y - k)^2 = r^2$
Substitute known values.	$(x - 3)^2 + (y - 1)^2 = 5^2$
Simplify the equation.	$(x - 3)^2 + (y - 1)^2 = 25$

Ask "Does $(7, 4)$ satisfy the equation?"	$(7 - 3)^2 + (4 - 1)^2 \overset{?}{=} 25$
Simplify the equation.	$4^2 + 3^2 \overset{?}{=} 25$
	$16 + 9 \overset{?}{=} 25$
	$25 = 25$ ✔

The point $(7, 4)$ satisfies the equation and, therefore, lies on the circle.

Write the equation for each circle with the given center and radius. Then determine whether the test point lies on the circle.

1. $r = 13$, $C(-1, 2)$, test point $(-5, 14)$ _____

2. $r = 10$, $C(4, -2)$, test point $(10, -10)$ _____

3. $r = 9$, $C(0, 3)$, test point $(-4, -5)$ _____

4. $r = 17$, $C(-5, -3)$, test point $(10, 5)$ _____

5. $r = 8$, $C(-4, 2)$, test point $(2, -3)$ _____

6. $r = 5$, $C(1, 1)$, test point $(-3, -2)$ _____

7. **Critical Thinking** The method described above can also be used to determine whether a particular point lies inside of a circle or outside of a circle. How could you determine this by using the equation of a circle and a test point?

_____

_____

_____

Problem Solving/Critical Thinking

12.6 *The Tangent Function*

In order to evaluate trigonometric functions like the tangent function, you must be able to identify the adjacent and the opposite sides of a right triangle. When you are working with two angles in a right triangle, the opposite and adjacent sides will be different for each angle.

◆ **Example**
Find the tangent of ∠A in the triangle shown at right.

◆ **Solution**
First, label the sides of the triangle to show which side is opposite and which side is adjacent.

Then, use the values to find the tangent ratio.

$$\tan A = \frac{\text{opposite}}{\text{adjacent}} = \frac{6}{8} = \frac{3}{4}$$

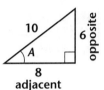

Label the opposite and the adjacent sides for each labeled angle.

1.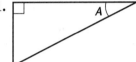

2.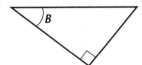

3.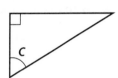

Find the tangent ratio for both labeled angles in each triangle.

4.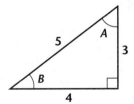

5.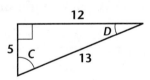

6.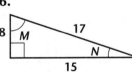

tan A = _____

tan B = _____

tan C = _____

tan D = _____

tan M = _____

tan N = _____

7. **Critical Thinking** If angles *A* and *B* are acute angles in a right triangle, how are tan *A* and tan *B* related?

_____

_____

Problem Solving/Critical Thinking

12.7 *Labeling Diagrams to Evaluate The Sine and Cosine Functions*

In order to evaluate trigonometric functions, such as sine and cosine, you must be able to identify the adjacent and the opposite sides of a right triangle. When you are working with two angles in a right triangle, the opposite and adjacent sides will be different for each angle.

◆ **Example**
   Find the sine and cosine of $\angle A$ in the triangle shown at right.

◆ **Solution**
   First, label the sides of the triangle to show which side is opposite and which side is adjacent.

   Then, use the values to find the sine and cosine ratios.

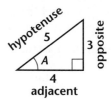

$$\sin A = \frac{\text{opposite}}{\text{hypotenuse}} = \frac{3}{5} \qquad \cos A = \frac{\text{adjacent}}{\text{hypotenuse}} = \frac{4}{5}$$

Label the opposite and adjacent sides for each labeled angle.

1.

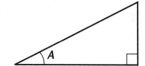

2.

3.

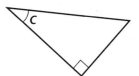

Find the sine and cosine ratios for the labeled angle in each triangle.

4.

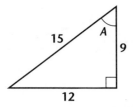

5.

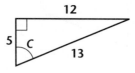

6.

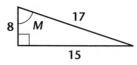

$\sin A =$ _____ $\sin C =$ _____ $\sin M =$ _____

$\cos A =$ _____ $\cos C =$ _____ $\cos M =$ _____

7. **Critical Thinking** If angles A and B are acute angles in a right triangle, how are $\sin A$ and $\cos B$ related?

_____

_____

_____

Problem Solving/Critical Thinking

12.8 *Colored Diagrams for Matrix Computation*

You will need colored pencils for this worksheet.

When you are adding or subtracting matrices, it can be difficult to keep track of the entries. Try using colored pencils to help. Here are some suggested strategies.

$$\begin{bmatrix} \boxed{38} & 62 & \boxed{59} \\ 84 & \boxed{63} & 75 \\ \boxed{59} & 42 & \boxed{71} \end{bmatrix}$$

- Shade the entries in the first matrix red and the entries in the second matrix blue.

- Shade the first column of each matrix red; the second column blue.

- Shade five entries in each matrix as shown by the boxes at the right. This makes a checkerboard pattern.

Use colored pencils to help you with these problems. Use one of the strategies suggested above or invent a strategy of your own.

1. $\begin{bmatrix} 38 & 62 & 59 \\ 84 & 63 & 75 \\ 59 & 42 & 71 \end{bmatrix} + \begin{bmatrix} 23 & 41 & 60 \\ 43 & 51 & 70 \\ 92 & 43 & 55 \end{bmatrix} =$

2. $\begin{bmatrix} 39 & 26 & 41 \\ 62 & 53 & 68 \\ 42 & 36 & 29 \end{bmatrix} - \begin{bmatrix} 17 & 19 & 31 \\ 43 & 28 & 34 \\ 25 & 17 & 12 \end{bmatrix} =$

3. $\begin{bmatrix} 458 & 370 & 210 \\ 435 & 392 & 371 \\ 418 & 420 & 205 \end{bmatrix} - \begin{bmatrix} 371 & 280 & 195 \\ 266 & 143 & 277 \\ 320 & 281 & 149 \end{bmatrix} =$

4. $\begin{bmatrix} 342 & 286 & 195 \\ 204 & 465 & 381 \\ 144 & 297 & 365 \end{bmatrix} + \begin{bmatrix} 181 & 392 & 168 \\ 240 & 351 & 478 \\ 226 & 193 & 354 \end{bmatrix} =$

5. $\begin{bmatrix} -7 & 12 & 9 \\ -6 & -5 & 4 \\ -7 & 11 & -8 \end{bmatrix} - \begin{bmatrix} -3 & 4 & -5 \\ -6 & 7 & -9 \\ -4 & 3 & -8 \end{bmatrix} =$

6. $\begin{bmatrix} -3 & 2 & -1 \\ 0 & -2 & 3 \\ -1 & 0 & -2 \end{bmatrix} + \begin{bmatrix} -1 & -2 & 0 \\ 0 & 2 & 3 \\ -3 & -2 & 1 \end{bmatrix} =$

7. **Critical Thinking** Addition of matrices is both commutative and associative. Explain how applying these properties and how using the results of Exercise 6 can help you simplify this problem.

$\begin{bmatrix} -1 & -2 & 0 \\ 0 & 2 & 3 \\ -3 & -2 & 1 \end{bmatrix} - \begin{bmatrix} 2 & -1 & 3 \\ -2 & 1 & -3 \\ 2 & -1 & 0 \end{bmatrix} + \begin{bmatrix} -3 & 2 & -1 \\ 0 & -2 & 3 \\ -1 & 0 & -2 \end{bmatrix} =$

Problem Solving/Critical Thinking

13.1 *Diagrams and Geometric Probability*

Diagrams can visually represent probabilities based on the ratio of areas within geometric figures. To determine the probability of an event occurring within two geometric figures, divide the area of one figure, where the event is successful, by the area of the surrounding figure.

◆ **Example**
The pattern on the dartboard shown is an 8-inch square within a 24-inch-by-30-inch rectangle. Kinesha throws a dart at the board. What is the probability that it lands within the square?

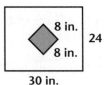

◆ **Solution**
$$P(\text{dart lands in square}) = \frac{\text{area of square}}{\text{area of rectangle}} = \frac{8 \times 8}{24 \times 30} = \frac{64}{720}, \text{ or } \frac{4}{45}$$

The probability that the dart lands within the square is $\frac{4}{45}$, or about 8.9%.

A dart lands on a random point within each target shown. Find the probability that it lands in the shaded area.

1.

10 in.

10 in.

2.

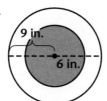

9 in.

6 in.

3.

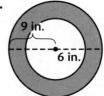

9 in.

6 in.

_____   _____   _____

A dart lands on a random point within each target shown. The player is a winner if the dart lands in the shaded region of the target. Give the dimensions of the shaded region that will result in a 25% probability of winning when one dart is thrown.

4.

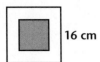

16 cm

16 cm

5.

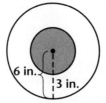

6 in.

3 in.

6.

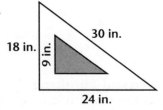

30 in.

18 in.

9 in.

24 in.

_____   _____   _____

Problem Solving/Critical Thinking

13.2 *Drawing Venn Diagrams to Determine Probability*

Drawing a Venn diagram is an excellent way to visualize probability problems. Once you have drawn the diagram, you can easily determine the probability of various combinations of events by using the Addition Counting Principle.

◆ **Example**

In a certain class, there are 10 girls and 15 boys. Five of the girls and 7 of the boys live on the east side of town; 3 girls and 6 boys live on the west side of town, and the rest of the students live in the country.

Draw a Venn diagram that illustrates the situation.

◆ **Solution**

Starting by drawing two separate circles to represent boys and girls because there is no intersection for these sets.

Then draw three ovals to represent each area where students live. Each of these should intersect with both the circles to show that boys and girls live in each area.

Finally, add numbers to show how many students are in each section of your diagram.

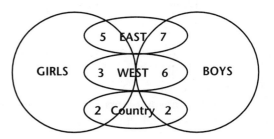

Draw a Venn diagram to illustrate each situation.

1. On a basketball team, there are 20 players. Eleven of the players are seniors; 10 of the players are over 6 feet, and 7 of the players play football. Three of the seniors play football, but only 1 of these is over 6 feet. Three of the other seniors are over 6 feet tall and do not play football. Only 2 of the football players are over 6 feet tall.

2. The following table summarizes the data from the recent student council election. Votes for the third candidate were not recorded.

	Boys	Girls
Voted for Student Council	142	156
Voted for Kari King	48	71
Voted for Stan Silver	64	60

3. **Critical Thinking** How many boys and girls voted for the third candidate?

_____

Problem Solving/Critical Thinking

13.3 Choosing the Best Method of Solution: Probability

When you are working different types of probability problems, you must decide if you need to use the Addition Counting Principle, the Fundamental Counting Principle, or neither.

◆ **Example**
In a certain class, there are 10 girls and 15 boys. Five of the girls and 7 of the boys live on the east side of town; 3 girls and 6 boys live on the west side of town, and the rest of the students live in the country.

You want to know the probability that if one student is randomly chosen to represent the class, he or she will live in town. Would you use the Addition Counting Principle, the Fundamental Counting Principle, or neither?

◆ **Solution**
This probability involves a union, so you will use the Addition Counting Principle.

◆ **Example**
If you want to know the probability of drawing 2 aces in a row from a deck of playing cards with replacement, which principle, if any, would you use?

◆ **Solution**
This probability involves a first and second choice, so you will use the Fundamental Counting Principle.

For each situation, determine whether you would use the Addition Counting Principle, the Fundamental Counting Principle, or neither.

1. The probability of rolling double 6s on a pair of 6-sided number cubes.

_____

2. The probability of drawing the two of hearts out of a deck of cards with replacement.

_____

3. The probability that a random order is either a hamburger or a chicken sandwich if 6 members of the team ordered hamburgers, 5 members of the team ordered chicken sandwiches, and 9 members of the team ordered tacos.

_____

4. If the chance of rain is 50% all week, the probability that it rains 3 days in a row.

_____

5. Critical Thinking For Exercise 4, how would the probability be affected if the problem were changed to reflect the chance that it rains 4 days in a row?

_____

Problem Solving/Critical Thinking
13.4 Drawing Area Models and Tree Diagrams

Diagrams and area models are effective tools for solving probability problems. You can use both area models and tree diagrams to make problems easier to visualize and to solve.

◆ **Example**

For the end-of-the-year banquet, students may choose one main course and one dessert from the following menu:

Main course: meatloaf, grilled chicken, fish filet, or vegetable lasagna
Desserts: chocolate cake or cherry pie

Draw an area diagram and a tree diagram to illustrate the situation. Assume that all choices are equally likely.

◆ **Solution**

For the area model, draw a square and divide it horizontally into four equal sections. Then divide it vertically into two sections. There should be eight sections to represent each of the possible combinations.

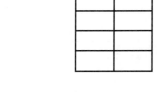

For the tree diagram, the first branch should show four choices, and the second branch should show two choices. There should be 8 final choices to represent the possible combinations.

Draw an area model and a tree diagram to illustrate each situation. Assume that all choices are equally likely.

1. The number of possible choices when choosing a jacket and a hat out of a selection of 3 jackets and 2 hats.

2. The number of possible choices when buying 2 gumballs from a gumball machine that has equal numbers of red, blue, green, and yellow gumballs.

3. The number of possible outcomes when flipping a coin and choosing a number between 1 and 5.

4. The number of possible outcomes when rolling a pair of 6-sided number cubes.

5. **Critical Thinking** When you roll a pair of 6-sided number cubes, why is the probability of rolling a sum of a seven higher than the probability of any other sum?

Problem Solving/Critical Thinking

13.5 *Comparing Experimental and Theoretical Probabilities*

Conducting simulations to find experimental probabilities is an excellent way to test theoretical probabilities.

◆ **Example**
Design and run a simulation to test the theoretical probability of rolling a 1 or a 2 on a 6-sided number cube.

◆ **Solution**
First, calculate the theoretical probability. the possibility of rolling a 1 is $\frac{1}{6}$, and the possibility of rolling a 2 is $\frac{1}{6}$, so the theoretical probability of rolling either number is $\frac{1}{6} + \frac{1}{6} = \frac{2}{6}$, or $\frac{1}{3}$.

Next run a simulation. The table shows the results of rolling a number cube 15 times.

Trail	1	2	3	4	5	6	7	8	9	10	11	12	13	14	15
Result	1	6	4	1	3	2	3	4	2	1	6	1	6	6	3

The number 1 was rolled 4 times and the number 2 was rolled twice.

The experimental probability was $\frac{6}{15}$, or $\frac{2}{5}$. This probability, 40%, is slightly higher than the theoretical probability, $33\frac{1}{3}$%.

Calculate the theoretical probability for each event. Then perform a simulation with 20 trials to find the experimental probability for the same event.

1. The probability of drawing a diamond out of a fresh deck of 52 cards.

_____

2. The probability of rolling 7 on a pair of 6-sided number cubes.

_____

3. The probability that 2 coins will match when they are flipped.

_____

4. **Critical Thinking** As you continue to increase the number of trials, how will the experimental probability of an event compare to the theoretical probability of the same event?

_____

Problem Solving/Critical Thinking

14.1 Using Mapping Diagrams

Drawing mapping diagrams is an effective way to determine whether a given relation, or set of ordered pairs, represents a function. When each element in the domain is paired with exactly one element in the range, the relation is a function.

◆ **Example**
Draw a mapping diagram to determine whether the given relation represents a function.

$$\{(1, 9), (2, 6), (1, 6), (4, 6), (5, 9)\}$$

◆ **Solution**
Draw a diagram containing two ovals, one for the domain and one for the range.
Draw an arrow from 1 to 9. Draw an arrow from 2 to 6.
Draw an arrow from 1 to 6. Once you see two arrows from an element of the domain going to two elements in the range, you can stop. You know that the relation is not a function.

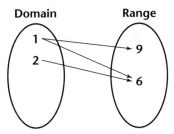

Draw a mapping diagram to determine whether each relation represents a function. Explain why or why not.

1. $\{(a, g), (b, h), (c, g), (d, e), (b, f)\}$

    _____

    _____

2. $\{(1.2, 3.8), (3.8, 6.5), (6.5, 8.4), (8.4, 9.1), (9.1, 12.3)\}$

    _____

    _____

3. $\{(\text{black}, 32A), (\text{gold}, 19C), (\text{red}, 32B), (\text{white}, 7D),$
    $(\text{gold}, 16B), (\text{red}, 28D)\}$

    _____

    _____

4. **Critical Thinking** Many relations are created using time measurements as elements in the domain and corresponding measurements as elements in the range. Why are these relations usually functions?

    _____

    _____

Problem Solving/Critical Thinking

14.2 *Using a Key Point to Identify Translations*

Suppose that you are asked to identify any translations of the parent function found in the graph of the function $y = (x - 2)^2 + 3$. You know what a translation means, and you know that there are horizontal translations and vertical translations. Unfortunately, you get confused about how the numbers in the equation tell you the direction of a translation.

One way to help correctly identify translations is to analyze the motion of a key point on the graph of the parent function. This is sufficient to identify the complete transformation.

◆ **Example**

Identify any translations of the parent function found in the graph of $y = (x - 2)^2 + 3$.

◆ **Solution**

Begin by identifying the parent function as $y = x^2$. Choose a key point on the graph. In this case, the vertex $(0, 0)$ is the key point.

Next, identify the coordinates of the vertex of the transformed function. The graph of $y = (x - 2)^2 + 3$ has a vertex with the coordinates $(2, 3)$.

Finally, analyze the movement from $(0, 0)$ to $(2, 3)$ via translations. The analysis at right indicates that the vertex moves 2 units to the right and three units up. The graph of $y = (x - 2)^2 + 3$ is a horizonal translation of the parent function 2 units to the right and a vertical translation 3 units up.

> x-coordinate: $0 \longrightarrow 2 \longrightarrow$ Add 2. (Right 2)
>
> y-coordinate: $0 \longrightarrow 3 \longrightarrow$ Add 3. (Up 3.)

For each function, identify the coordinates of the vertex. Use these and the key point for its parent function to identify each transformation.

1. $y = (x + 3)^2 + 5$ _____

2. $y = (x - 2)^2 - 3$ _____

3. $y = (x + 6)^2 - 6$ _____

4. $y = (x - 1.5)^2 + 4$ _____

5. $y = |x + 2.5| - 4$ _____

6. $y = |x - 2| - 2.4$ _____

7. $y = |x + 3.8| + 4$ _____

8. $y = |x - 100| + 100$ _____

9. **Critical Thinking** Let $y = 4(2^x) + 1$. The graph of this function is both a horizontal and a vertical translation of the parent function. Identify the translations and justify your response.

_____

_____

Problem Solving/Critical Thinking

14.3 *Generalizing Positive Scale Factors*

When working with functions that involve stretches, the scale factor can provide many clues about the functions and their graphs.

You can use squares and rectangles to examine how different positive scale factors affect the graph of various functions.

◆ **Example**
Describe the effects of the scale factor of 3 in the function $y = 3x^2$. Describe how the factor affects the y-values of the function at each x-value and how the factor affects the graph.

◆ **Solution**
If $x = 2$, then x^2 has a value of 4. If $x = -2$, x^2 has a value of 4 also. You can see the graph of $y = x^2$ and the square whose vertices are $(2, 0)$, $(-2, 4)$, $(2, 4)$, and $(2, 0)$ in the diagram at right. If $x = 2$, then $3x^2$ has a value of 12. If $x = -2$, $3x^2$ has a value of 12 also. In the diagram, you can also see the graph of $y = 3x^2$ and the rectangle whose vertices are $(-2, 0)$, $(-2, 12)$, $(2, 12)$, and $(2, 0)$. Notice that the scale factor of 3 has the effect of vertically stretching the square into a rectangle. That is, the square becomes a narrower rectangle. The effect on the parabola is the same, that is, the graph of $y = 3x^2$ is narrower than the graph of $y = x^2$.

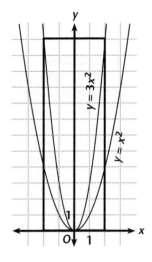

Describe how the scale factor affects the y-values of each function for the given x-values and how the factor affects the graph.

1. $y = 9|x|$ _____

2. $y = 0.55|x|$ _____

3. $y = 3.75|x|$ _____

4. $y = 6.9x^2$ _____

5. $y = \frac{1}{3}x^2$ _____

6. **Critical Thinking** What generalization can you make about the effects of a scale factor that is *greater than one* on quadratic and absolute value functions?

_____

_____

7. **Critical Thinking** What generalization can you make about the effects of a scale factor that is *less than one* on quadratic and absolute value functions?

_____

_____

Problem Solving/Critical Thinking

14.4 Generalizing Negative Scale Factors

When working with functions that involve reflections and stretches, the scale factor will be a negative number.

You can use squares and rectangles to examine how different negative scale factors affect the graphs of various functions.

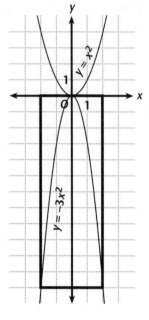

◆ **Example**

Describe how the scale factor of -3 affects the y-values of the function $y = -3x^2$ at each x-value and how the factor affects the graph.

◆ **Solution**

If $x = 2$, then x^2 has a value of 4. If $x = -2$, x^2 has a value of 4 also. You can see the graph of $y = x^2$ and the square whose vertices are $(2, 0)$, $(-2, 4)$, $(2, 4)$, and $(2, 0)$ in the diagram at right. If $x = 2$, $-3x^2$ has a value of -12. If $x = -2$, $-3x^2$ has a value of -12 also. In the diagram, you can also see the graph of $y = -3x^2$ and the rectangle whose vertices are $(-2, 0)$, $(-2, -12)$, $(2, -12)$, and $(2, 0)$. Notice that the scale factor of -3 has the effect of vertically stretching and flipping the square into a rectangle. That is, the square becomes a narrower flipped rectangle. The effect on the parabola is the same, that is, the graph of $y = -3x^2$ is narrower than that of $y = x^2$ and is flipped.

Describe how the scale factor affects the y-values of each function for given x-values and how the factor affects the graph.

1. $y = -0.25|x|$ _____

2. $y = -14|x|$ _____

3. $y = -8.1x^2$ _____

4. $y = -\frac{2}{7}x^2$ _____

5. **Critical Thinking** What generalization can you make about the effects of a scale factor that is *less than zero but greater than negative one* on quadratic and absolute value functions?

   _____

6. **Critical Thinking** What generalization can you make about the effects of a scale factor that is *less than negative one* on quadratic and absolute value functions?

   _____

Problem Solving/Critical Thinking

14.5 *Algebraic Tests for Symmetry*

An equation is symmetric about the
- **y-axis** when x is replaced by $-x$, and the equation remains unchanged.
- **x-axis** when y is replaced by $-y$, and the equation remains unchanged.
- **origin** when x is replaced by $-x$ and y is replaced by $-y$, and the equation remains unchanged.
- **line** $y = x$ when x is interchanged with y, and the equation remains unchanged.

◆ **Example**
Examine the equation $xy = 4$ to determine whether its graph has symmetry.
Draw a graph.

◆ **Solution**
Perform each symmetry test.

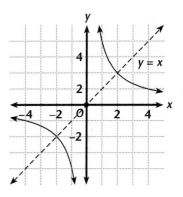

Substitute $-x$ for x.
$$xy = 4$$
$$(-x)y = 4$$
$$-xy = 4$$
The equation changes.

Substitute $-y$ for y.
$$xy = 4$$
$$x(-y) = 4$$
$$-xy = 4$$
The equation changes.

Substitute $-x$ for x
and $-y$ for y.
$$xy = 4$$
$$(-x)(-y) = 4$$
$$xy = 4$$
The equation is unchanged.

Interchange x and y.

$$xy = 4$$
$$yx = 4$$
$$xy = 4$$
The equation is unchanged.

Examine each equation algebraically to determine whether it has symmetry. If there is symmetry, name the lines or point of symmetry. Draw a graph to verify your finding.

Equation	**Symmetry**	**Graph**

1. $y = x^2$ _____

2. $x = y^2$ _____

3. $x^2 + y^2 = 25$ _____

4. $y = x^3$ _____

5. $y = x^4$ _____

6. $x^3 - y^2 = 0$ _____

ANSWERS

Problem Solving/ Critical Thinking—Chapter 1

Lesson 1.1

1. first row: 7
   second row: 53, 29, 11
   third row: 30, 24, 18
   fourth row: 6, 6

2. first row: 13
   second row: 24, 15
   third row: 15, 12, 9
   fourth row: 3, 3

3. 72, 56, 42   **4.** 29, 14, 3

5. 114, 55, 20   **6.** 78, 30, 4

7. Each number in Pascal's triangle is the sum of the two numbers above it, which is the reverse of the method of constant differences.

Lesson 1.2

1. Answers may vary. Sample answer: Choose a negative number such as -5 so the left side of the equation will be less than 54.

2. Answers may vary. Sample answer: Choose a positive multiple of 5, such as 50, so that the expression $\frac{4}{5}x$ becomes a whole number.

3. Answers may vary. Sample answer: Choose a positive fraction with a numerator of 3. Any number that is 3 or greater will make the fraction $\frac{3}{x}$ too small.

4. Answers may vary. Sample answer: Choose a negative number so that the left side will be positive.

5. too small   **6.** too large

7. too large   **8.** too small

9. Answers may vary. Sample answer: You can find out if the solution should be positive or negative, a fraction or a whole number, and whether the next guess should be larger or smaller.

Lesson 1.3

Answers may vary. Sample answers:

1. Evaluate 8^2; multiply $3 \cdot 24$.

2. Subtract $(6 - 3)$; add $(8 + 5)$.

3. Subtract $(27 - 6)$; multiply $(8)(26)$.

4. Multiply $4 \cdot 2$; add the result to 73.

5. Evaluate 7^2; subtract the result from 82.

6. Subtract $(7 - 2)$; subtract $(8 - 4)$.

7. Step 1: Evaluate $(16 - 9)(23)$. First, subtract 9 from 16. Multiply the answer by 23.
   Step 2: Evaluate $3^5 \cdot (6 + 4 \cdot 8)$. Multiply 4 times 8 and then add 6 to the product. Multiply this by the value of 3^5.
   Step 3: Subtract the results of Step 2 from the results of Step 1.

8. Step 1: Evaluate $5 + 4^3 - 11 \cdot 2$. Compute 4^3. Add this to 5. Subtract the product of 11 and 2.
   Step 2: Evaluate $(51 + 6)(3 \cdot 2 + 4)$. Add $51 + 6$. Multiply $3 \cdot 2$ and add 4 to the product. Multiply this by the sum $51 + 6$.
   Step 3: Divide the results of Step 1 by the results of Step 2.

9. **a.** Compute 2^3; then multiply $9 \cdot 4$. Subtract $36 - 28 - 8$ from left to right. The value of the numerator is 0.
   b. It is clear that the denominator does not equal 0. Since the numerator equals 0, the entire expression must also equal 0.

Lesson 1.4

1. **a.** The temperature goes down, stays constant, and then goes down again.
   b. at the beginning of the time period shown

2. **a.** The cost increases until a certain point. Then it levels off and stays constant.
   b. The number of pounds that you can ship goes down.

ANSWERS

3. a. The profits go down, stay the same, and then start to go up.
b. Assuming that both axes show positive numbers, the store has been making a profit during the entire time shown on the graph.

4. Answers may vary. Sample answer: The line on the graph moves up and down in an unpredictable manner.

Lesson 1.5

1. First differences: 10
Equation: $y = 10x + 60$
Sample situation: A fitness club charges a $60 initiation fee plus $10 per month.
2. First differences: 5
Equation: $y = 5x + 12$
Sample situation: A cassette club charges a $12 membership fee plus $5 per cassette purchased.
3. First differences: 8
Equation: $y = 8x$
Sample situation: A store sells pictures for $8 each.
4. First differences: 3
Equation: $y = 3x$
Sample situation: Ben has a collection of model cars that weigh 3 pounds each.
5. First differences: 1
Equation: $y = x + 8$
Sample situation: A carnival charges $8 for admission plus $1 per ride.
6. Sample answer: A bowl weighing 8 ounces is filled with candies weighing 1 ounce each, where x is the number of candies and y is the total weight in ounces.

Lesson 1.6

1. negative correlation **2.** no correlation

3. negative correlation

4. You would reach the same conclusion. When the science scores are put in increasing order the math scores, in general, increase. There is a positive correlation between the two sets of data.

Problem Solving/ Critical Thinking—Chapter 2

Lesson 2.1

1.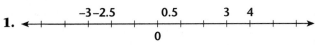

2.

3.

4.

5. $-\frac{5}{2}, -1\frac{1}{2}, \frac{1}{2}, 3, 4$

6. $-|x + 4|, -|x + 3|, -|x - 2|,$ $|x + 4|, |x + 5|$

Lesson 2.2

1. $-2 + 4 = 2$ **2.** $4 + (-5) = -1$

3. $0 + (-4) = -4$ **4.** $-9 + 3 = -6$

5. -5 **6.** -2 **7.** 3 **8.** 3

9. Answers may vary. Sample answer: Number the line from -60 to 90. Then use it to find the sum of $-50 + 80$. Add this sum to the sum of $-3 + 7$ for the final answer.

Lesson 2.3

1. $-2 + (-3) = -5$ **2.** $4 + 2 = 6$

3. $-3 + 5 = 2$ **4.** $-6 + 6 = 0$

5. $-8 + (-2) = -10$

6. Answers may vary. Sample answer: The student's drawing has 10 at the top and –10 at the bottom. To add a positive integer, move up. To add a negative integer, move down. To subtract, rewrite the problem as addition.

ANSWERS

Lesson 2.4

1.

x	−4	−3	−2	−1	0	1	2	3	4
−4	16	12	8	4	0	−4	−8	−12	−16
−3	12	9	6	3	0	−3	−6	−9	−12
−2	8	6	4	2	0	−2	−4	−6	−8
−1	4	3	2	1	0	−1	−2	−3	−4
0	0	0	0	0	0	0	0	0	0
1	−4	−3	−2	−1	0	1	2	3	4
2	−8	−6	−4	−2	0	2	4	6	8
3	−12	−9	−6	−3	0	3	6	9	12
4	−16	−12	−8	−4	0	4	8	12	16

2. Answers may vary. Sample answer: Students might shade the upper left and lower right quadrants red and then shade the upper right and lower left quadrants blue.

3. a. upper left quadrant
   b. lower right quadrant
   c. upper left and lower right quadrants
   d. upper right and lower left quadrants

4. The products in the upper left and lower right quadrants are positive, so multiplying two integers with like signs results in a positive product. The products in the upper right and lower left quadrants are negative, so multiplying two integers with unlike signs results in a negative product.

5. Answers may vary. Sample answer: Students could add two rows to the top of the table to include −6 in the left column. The product is −18.

6. Answers may vary. Sample answer: Find the dividend in the table. Find the divisor in either the top row or the left column. The quotient is in the top row if the divisor is in the left column. The quotient is in the left column if the divisor is in the top row.

Lesson 2.5

1. $\boxed{4-4}+2+2$

2. $\boxed{-5+5}+3+3+2$

3. $\boxed{-4+4}+6-2$

4. $\boxed{-2+2}+\boxed{3+4-5-2}$

5. $\boxed{-4+4}-4-4+14$

6. $\boxed{5\cdot\frac{1}{5}}\cdot\frac{2}{5}\cdot 4$

7. $\boxed{\frac{1}{2}\cdot 2}\cdot\frac{1}{4}\cdot 5$

8. $\boxed{8\cdot\frac{1}{8}}\cdot\frac{1}{2}\cdot 4$

9. $\boxed{\frac{4}{5}\cdot\frac{5}{4}}\cdot\frac{2}{5}\cdot\frac{3}{2}\cdot\frac{3}{5}$

10. $\boxed{2\cdot\frac{1}{2}}\cdot\boxed{\frac{2}{3}\cdot\frac{3}{2}}\cdot 3$

11. $\boxed{\frac{7}{8}\cdot\frac{8}{7}}\cdot\frac{1}{7}\cdot\frac{9}{8}\cdot\frac{1}{8}$

12. $\boxed{\frac{1}{4}\cdot 4}\cdot\boxed{5\cdot\frac{1}{5}}\cdot 2\cdot\frac{1}{10}\cdot\frac{3}{5}\cdot 3$

13. $\boxed{\frac{4}{5}\cdot\frac{5}{4}}\cdot\boxed{\frac{2}{5}\cdot\frac{5}{2}}\cdot\frac{1}{4}\cdot\frac{3}{2}\cdot\frac{6}{5}\cdot\frac{1}{2}\cdot\frac{3}{5}$

14. Answers may vary. Sample answer: The next steps could be $60+2-80+1 = -20+3 = -17$. This method might help someone use mental math to do the problem.

Lesson 2.6

1. $(-3x-4)+(-2+x)$; $-2x-6$

2. $(-1+b)+(-2b+4)+(-1+3b)$; $2b+2$

3. $(p+3)+(4-2q)+(-4q+2p)$; $3p-6q+7$

4. $(-c+2d)+(-d-1)+(3c-2d)+(-3-4c)$; $-2c-d-4$

ANSWERS

5. Answers may vary. Sample answer:

				sum
c-terms	$-c$	$3c$	$-4c$	$-2c$
d-terms	$2d$	$-d$	$-2d$	$-d$
constants	-1	-3		-4

Lesson 2.7

1. $-x - 4$　　**2.** $-6w^2 + 2$　　**3.** $-3 + y^2$

4. $-3a + 4$　　**5.** $-2 - 3p$　　**6.** $d^2 - 3$

7. $-x^2 + 2x - 1$　　**8.** $-2y^2 + 4y - 10$

9. $-2b^2 - b + 1$　　**10.** $2n^2 - n + 4$

11. $(3x^2 - 6x + 9) \div \dfrac{1}{3x}$

$= (3x^2 - 6x + 9) \cdot 3x$

$= 9x^3 - 18x^2 + 27x$

Problem Solving/ Critical Thinking—Chapter 3

Lesson 3.1

1. a whole number, greater than 9

2. a decimal, less than 8

3. a fraction greater than 1, a positive number

4. an integer, a negative number

5. a decimal, a negative number

6. a positive number, greater than 12

7. a negative fraction, less than $\dfrac{-1}{3}$

8. a negative integer, less than -7

9. a. Answers may vary. Sample answers: The value of n will probably be a positive number and a decimal greater than 20.
 b. Answers may vary. Sample answers: The value of h will probably be a positive number and a fraction greater than $\dfrac{1}{2}$.

Lesson 3.2

1. a whole number, greater than 8

2. a fraction, less than 13

3. a decimal, greater than 5

4. a negative number, greater than -2.4

5. a decimal, less than 0.2

6. a negative number, less than $-\dfrac{4}{5}$

7. Answers may vary. Sample answer: The solution will probably be a negative number greater than -5.

Lesson 3.3

1. a. $\dfrac{x}{5} + 7 - 7 = -2 - 7$

$\dfrac{x}{5} = -9$

$\dfrac{x}{5} \cdot 5 = -9 \cdot 5$

$x = -45$

 b. $5 \cdot \dfrac{x}{5} + 5 \cdot 7 = 5 \cdot -2$

$x + 35 = -10$

$x + 35 - 35 = -10 - 35$

$x = -45$

2. a. $3x - 9 + 9 = 27 + 9$

$3x = 36$

$3x \div 3 = 36 \div 3$

$x = 12$

 b. $\dfrac{3x}{3} - \dfrac{9}{3} = \dfrac{27}{3}$

$x - 3 = 9$

$x - 3 + 3 = 9 + 3$

$x = 12$

3. a. $\dfrac{x}{2} - 4 + 4 = 3 + 4$

$\dfrac{x}{2} = 7$

$\dfrac{x}{2} \cdot 2 = 7 \cdot 2$

$x = 14$

 b. $2 \cdot \dfrac{x}{2} - 2 \cdot 4 = 2 \cdot 3$

$x - 8 = 6$

$x - 8 + 8 = 6 + 8$

$x = 14$

4. a. $10 + 5x - 10 = 20 - 10$
$5x = 10$
$5x \div 5 = 10 \div 5$
$x = 2$

b. $\dfrac{10}{5} + \dfrac{5x}{5} = \dfrac{20}{5}$
$2 + x = 4$
$2 + x - 2 = 4 - 2$
$x = 2$

5. Answers may vary. Sample answer: Choosing addition or subtraction first will often make the computation easier.

Lesson 3.4

1. $x = 4$

2. $a = 5$

3. $n = 2$

4. $s = -6$

5. $x = 84$

6. $z = -75$

7. Answers may vary. Sample answer: First clear the fractions, if any, move all the x-terms to the left side and all constant terms to the right side, and then divide by any remaining coefficient of x.

Lesson 3.5

1. $6x + 15 = 6x - 8$; no real solutions

2. $7p - 3 = 4p + 18$; one real solution

3. $-6k - 6 = -6k - 6$; all real numbers

4. $9x + 20 = 9x + 6$; no real solutions

5. $-2y - 10 = -2y - 10$; all real numbers

6. $2m + 12 = -2m + 12$; one real solution

7. $3.6z - 21.6 = -3.6z + 7.2$; one real solution

8. $12c - 9.6 = 12c - 5$; no real solutions

9. Answers may vary. Sample answer: If you subtract the x-term from each side of the equation, you get an equivalent equation stating that two different constants are equal, which is false.

Lesson 3.6

1. $k = m - l$ **2.** $h = d - j$

3. $x = \dfrac{z - 2y}{3}$ **4.** $x = \dfrac{4y + 3z}{5}$

5. $m = \dfrac{f}{a}$ **6.** $t = \dfrac{I}{pr}$

7. $g = \dfrac{w - h}{f}$ **8.** $f = \dfrac{d - g}{e}$

9. $x = \dfrac{y - 8z}{5}$ **10.** $x = \dfrac{4z - y}{3}$

11. $q = \dfrac{5p - r}{4}$ **12.** $r = \dfrac{7 - 2s}{6}$

13. Usually it means that a mistake has been made. You should recheck your work carefully and correct the mistake.

Problem Solving/ Critical Thinking—Chapter 4

Lesson 4.1

Answers may vary. Sample answers are given.

1. $\dfrac{13.4 \text{ miles}}{2.25 \text{ hours}} = \dfrac{26.2 \text{ miles}}{n \text{ hours}}$

2. $\dfrac{200 \text{ feet}}{3 \text{ yards}} = \dfrac{n \text{ feet}}{5.5 \text{ yards}}$

3. $\dfrac{1 \text{ liter}}{0.22702 \text{ gallons}} = \dfrac{n \text{ liters}}{40 \text{ gallons}}$

4. $\dfrac{4.2 \text{ hours}}{85 \text{ miles}} = \dfrac{n \text{ hours}}{145 \text{ miles}}$

5. $\dfrac{78 \text{ pages}}{2.25 \text{ hours}} = \dfrac{n \text{ pages}}{5 \text{ hours}}$

6. $\dfrac{525 \text{ dollars}}{1 \text{ month}} = \dfrac{n \text{ dollars}}{12 \text{ months}}$

ANSWERS

7. The supervisor did not use a proportion because the time needed for the job decreases as the number of people increases. A proportion such as $\dfrac{4 \text{ people}}{8 \text{ days}} = \dfrac{n \text{ people}}{2 \text{ days}}$ will not result in the answer that the supervisor found.

Lesson 4.2

Answers may vary. Sample answers are given.
1. 25% of 69 is 17.25. 23% of 60 is 13.8. The estimate is between 13 and 18.

2. 12.4 is 10% of 124. 11 is 11% of 100. The estimate is between 100 and 124.

3. 132 is 150% of 88. The estimate is a little more than 88.

4. 10% of 64 is 6.4. 20% of 64 is 12.8. 25% of 64 is 16. The estimate is between 20% and 25%.

5. 60 is 500% of 12. The estimate is a little more than 5%.

6. 10% of 146 is 14.6. 5% of 146 is 7.3. The estimate is a little more than 7.3.

7. 20% of 72 is 14.4. 120% of 72 is 86.4. The estimate is a little more than 87.

8. 82 is 100% of 82. The estimate is a little more than 82.

9. 4.2 is 4.2% of 100. 8.4 is 4.2% of 200. 16.8 is 4.2% of 400. The estimate is a little less than 400.

10. 0.28 is 40% of 0.7. The estimate is a little more than 40%.

11. The phrase uses the number 10 twice; many problems will not use numbers close to 10, 10%, or 100.

Lesson 4.3

Students should sketch a tree diagram for each problem.
1. 36   2. 4   3. 16   4. 12   5. 36

6. 4   7. 30   8. 16   9. 6   10. 100

11. 108 outcomes

Lesson 4.4

1. $92 + 87 + 90 + 93 + 87 = 449$; $449 \div 5 = 89.8$

2. Answers may vary. Sample answer: estimated mean = 91; differences: $+1, -4, -1, +2, -4$; total = -6; $-6 \div 5 = -1.2$; $91 + (-1.2) = 89.8$.

3. 75.6   4. 51   5. 36   6. 82.8   7. 124.8

8. Answers may vary. Sample answer: It would depend on the numbers but, with such a small set, it would probably be faster to just add the numbers and divide by 3.

Lesson 4.5

Answers may vary. Sample answers are given.
1. bar; names of the 10 people; heights by 1/4-ft increments from 5 ft through 8 ft

2. line; time in weeks showing 104 weeks; sales in dollar amounts numbered in multiples of $5,000

3. bar; number of cars from 0 through 10; number of households in multiples of 50

4. bar; names of different kinds of movies; number of people in multiples of 10

5. bar; income in intervals of $5000; number of airplane trips from 0 through 20

6. line; names of the 12 months; dollars in multiples of 100 or 1000, depending on the amounts in the savings account

7. bar; labels for the bars "with fertilizer" and "without fertilizer"; heights in multiples of 2 centimeters

8. line; dates for the first days of the weeks; temperature in multiples of 5 degrees

9. Answers may vary. Sample answer: If the graph must be very small, the axes should be numbered with multiples of a large number.

ANSWERS

Lesson 4.6

1. 23 people; both; stem-and-leaf is easier

2. 52; stem-and-leaf   3. 0; both   4. 2; both

5. cannot be determined; neither

6. 17 to 52; stem-and-leaf

7. 50 − 59; both; histogram is easier

8. cannot be determined; neither

9. 22 people; stem-and-leaf

10. an equal number; both; histogram is easier

11. 36; stem-and-leaf

12. 26 and 38; stem-and-leaf

13. Answers may vary. Sample answer: The histogram gives the information at a glance and is easier to read, but the stem-and-leaf plot has more information because it shows all the actual ages.

Problem Solving/ Critical Thinking—Chapter 5

Lesson 5.1

1. Not a function; the y-axis intersects the oval in two points.

2. function

3. Not a function; there are many vertical lines that intersect the curve in two points. One of them goes through (4, 0).

4. function

5. Not a function; the y-axis intersects the polygon in two points. The graph of a polygon cannot represent a function.

6. function

7. Once the curve goes straight up, it coincides with a vertical line, so the vertical line

intersects the curve in more than one point. Also, when the curve doubles back, it may move above or below itself. Vertical lines then would cut the curve in two points.

Lesson 5.2

1. right; below   2. left; above

3. right; above   4. right; below

5. right; below   6. left; above

7. left; above   8. left; above

9. Point B is to the left and above point A; negative slope.

10. Point B is to the right and below point A; negative slope.

11. Point B is to the right and above point A; positive slope.

12. **a.** a vertical line segment
    b. a horizontal line segment

Lesson 5.3

1. to change liters per minute to gallons per hour

2. to change ounces per day to pounds per week

3. to change centimeters per second to feet per minute

4. to change pounds per minute to grams per second

5. to change kilometers per week to meters per hour

6. to change millimeters per second to inches per minute

7. 0.1577 liters per second

8. 8858.27 feet per hour

9. 187.5 pounds per minute

10. 104.5 pounds per hour

ANSWERS

11. Answers may vary. Sample answer: Either conversion equation can be used, although most people prefer working with whole numbers rather than with decimals. To change pints into pecks, you could multiply by $\frac{1 \text{ peck}}{16 \text{ pints}}$.

Lesson 5.4

1. negative; less than −1

2. positive; less than 1

3. positive; equal to 1

4. positive; greater than 1

5. negative; equal to −1

6. negative; greater than −1

7. positive; less than 1

8. negative; greater than −1

9. Answers may vary. Sample answer: The line can have either a positive or a negative slope. The slope is rather flat.

Lesson 5.5

1. $-\frac{3}{2}; -\frac{3}{4}$ 2. $-\frac{1}{5}; -1$ 3. $2; 6$ 4. $-4; 2$

5. $-\frac{1}{3}; -\frac{1}{6}$ 6. $-8; -2$ 7. $6; 2$

8. $-1; -\frac{2}{3}$ 9. $-5; 5$ 10. $3; \frac{3}{2}$

11. $2; \frac{8}{3}$ 12. $-\frac{2}{3}; -2$

13. y-intercept = −3; x-intercept = 2

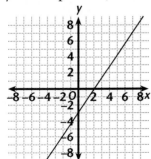

14. y-intercept = −4; x-intercept = −1

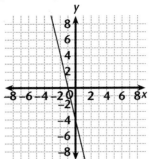

15. y-intercept = 2; x-intercept = −1

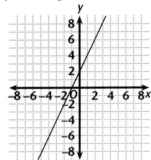

16. Answers may vary. Sample answer: Equations of the form $ay = b$ do not have an x-intercept because they represent horizontal lines. Equations of the form $ax = b$ do not have a y-intercept because they represent vertical lines.

Lesson 5.6

1. $2; 2$; parallel   2. $5; -\frac{1}{5}$; perpendicular

3. $2; -2$; neither   4. $1; -1$; perpendicular

5. $-\frac{1}{3}; -\frac{3}{2}$; neither   6. $-3; -3$; parallel

7. $-\frac{3}{4}; -\frac{3}{4}$; parallel   8. $-2; \frac{1}{2}$; perpendicular

9. $\frac{1}{4}; -4$; perpendicular

10. $-\frac{2}{3}; -\frac{3}{2}$; neither

11. Answers may vary. Sample answer: Horizontal lines will have slopes equal to 0; vertical lines will have slopes that are undefined. You can still use the slopes to determine if the lines are parallel, perpendicular, or neither.

ANSWERS

Problem Solving/ Critical Thinking—Chapter 6

Lesson 6.1

1. some of both   2. some of both

3. all negative   4. some of both

5. all positive   6. some of both

7. less than 16   8. greater than 5.7

9. greater than 4   10. less than -1.3

11. Answers may vary. Sample answer: The solution is a set of numbers greater than or less than some other number. This is a subset of the real numbers and, therefore, must contain both integers and fractions.

Lesson 6.2

1. $x < -1$ 2. $x > 9$ 3. $x > -7$

4. $x > -1.25$ 5. $x > 9$ 6. $x > -6$

7. $x < -20$ 8. $x < -1.3$ 9. $x > 22$

10. $x > -50$ 11. $x < 2$ 12. $x > \frac{1}{6}$

13. Solve the related equation. Then replace $=$ with either $\leq$ or $\geq$.

Lesson 6.3

1. ![number line with open circles at -4 and 3, shaded between]

2. ![number line with closed circles at -1 and 3, shaded between]

3. ![number line with closed circle at -4 and closed circle at 4, shaded between]

4. ![number line with closed circles at 2 and 4, shaded between]

5. ![number line with closed circles at -2 and 2]

6. ![number line with closed circles at -4 and 1]

7. ![number line with closed circles at -2 and 3]

8. ![number line with closed circles at -2 and 3]

9. Answers may vary. Sample answer:
   $-4 \leq x \leq 4$ and ($x \leq -1$ or $x \geq 1$).

Lesson 6.4

1.
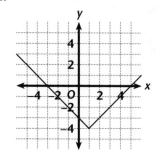

2.
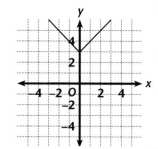

3.
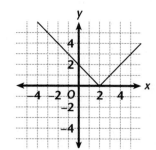

4.
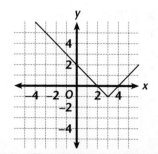

5.

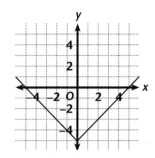

6.

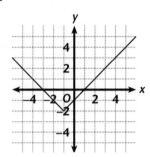

7.

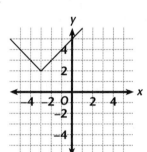

8.

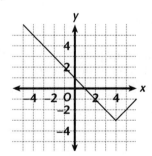

9.

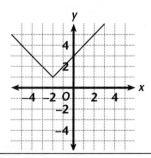

10. Answers may vary. Sample answer: h is subtracted because substituting h for x should result in a 0 inside the absolute value expression. k is added because adding a constant to y moves the graph up.

Lesson 6.5

1. $+(2x + 5) = 3,\ -(2x + 5) = 3$

2. $+(7 - x) = 2,\ -(7 - x) = 2$

3. $+(-3x - 5) = 8,\ -(-3x - 5) = 8$

4. $+(7 - 4x) > 1,\ -(7 - 4x) > 1$

5. $+(3 + 2x) \le 5,\ -(3 + 2x) \le 5$

6. $+(-x + 3) \ge 6,\ -(-x + 3) \ge 6$

7. $5 = +(5x + 2),\ 5 = -(5x + 2)$

8. $4 < +(6 - 2x),\ 4 < -(6 - 2x)$

9. $3 \ge +(4x + 1),\ 3 \ge -(4x + 1)$

10. no real numbers **11.** some real numbers

12. all real numbers **13.** some real numbers

14. all real numbers **15.** no real numbers

Problem Solving/ Critical Thinking—Chapter 7

Lesson 7.1

1. G **2.** C **3.** A **4.** F **5.** H

6. B **7.** I **8.** D **9.** 126

Lesson 7.2

1. $\begin{cases} 2x = 9 + 3y \\ 2(2x) - 51 = 5y \end{cases}$
$2(9 + 3y) - 51 = 5y;\ (54, 33)$

2. $\begin{cases} 3(3y) - 5 = -2x \\ 3y = -5x - 7 \end{cases}$
$3(-5x - 7) - 5 = -2x;\ (-2, 1)$

ANSWERS

- Algebra -

3. $\begin{cases} 3(5x) - 2y + 75 = 0 \\ 5x = 3y - 25 \end{cases}$
$3(3y - 25) - 2y + 75 = 0;\ (-5, 0)$

4. $\begin{cases} 4y = 3x - 6 \\ 2(4y) - 7x = -18 \end{cases}$
$2(3x - 6) - 7x = -18;\ (6, 3)$

5. $\begin{cases} 21 + 2(5x) = 3y \\ 5x = 2y - 9 \end{cases}$
$21 + 2(2y - 9) = 3y;\ (-3, -3)$

6. Answers may vary. Sample answer: The coefficients of either the x-terms or the y-terms must be multiples of each other.

Lesson 7.3

1. $y = 1$ **2.** $(2, 1)$ **3.** $(4, -1)$

4. $(2, 5)$ **5.** $(-3, -5)$ **6.** $(4, 3)$

7. Answers may vary. Check that students' charts are logical and will accommodate most steps in the solutions.

Lesson 7.4

1. $3x + y = 2$
$-2x + 5y = -10$
one solution

2. $4x - 3y = -2$
$12x - 9y = -6$
infinitely many solutions

3. $2x - 3y = -1$
$2x - 6y = -1$
one solution

4. $10x - 25y = 5$
$2x - 5y = 1$
infinitely many solutions

5. $-3x - 4y = 8$
$-5x + y = 2$
one solution

6. $6x - 9y = 10$
$2x - 3y = 5$
no solutions

7. $x - 7y = 2$
$3x - 21y = 6$
infinitely many solutions

8. $2x - y = 4$
$8x - 4y = -1$
no solutions

9. $4x - 2y = -1$
$8x - y = -2$
one solution

10. Answers may vary. Sample answer: First check the slopes. If the equations have different slopes, there is one solution. If the equations have the same slope and the same y-intercept, there are infinitely many solutions. If the equations have the same slope and different y-intercepts, there are no solutions.

Lesson 7.5

1. a number   **2.** an ordered pair

3. a number   **4.** a quantity

5. a word   **6.** a quantity

7. an equation   **8.** an equation

9. an equation, or a set of ordered pairs

10. a number

11. a set of ordered pairs

12. an ordered pair

13. a word   **14.** a number

15. Answers may vary. Sample answer: problems that ask you to draw a graph or make a geometric figure

Lesson 7.6

1. volume of each solution in liters

2. the number of kilometers she ran

3. number of hours until the trains are the same distance from Los Angeles

4. dollars collected for adults' tickets

5. number of dimes that Hal had

ANSWERS

6. gallons of skim milk

7. Answers may vary. Sample answer: Add the answer to 30 in order to find the number of gallons in the mixture. Find the number of gallons of butterfat in the 30-gallon barrel. Check to see if this amount equals 1.5% of the number of gallons in the mixture.

Problem Solving/ Critical Thinking—Chapter 8

Lesson 8.1

1. $-10x^4y^2$ **2.** $2a^2b^5c^3$ **3.** $24p^5q^5$

4. $15c^4d^2$ **5.** $4a^4b^3c^3d^3$ **6.** $-2w^3x^2y^3$

7. $3m^2n^3p^2$ **8.** $10w^3x^4y^2$ **9.** $-3g^5h$

10. $-5p^5q^5r^3$ **11.** $6hk^5m$ **12.** $-6a^3b^4c^2d^5$

13. Answers may vary. Sample answer: Count the number of different variables in all of the factors.

Lesson 8.2

1. power of a power

2. product of powers

3. power of a product

4. product of powers

5. power of a product

6. power of a power

7. power of a product

8. product of powers

9. power of a power

10. Product of Powers, then Power of a Power

11. Power of a Power twice, Product of Powers

12. Power of a Power, Power of a Product, Product of Powers

13. Power of a Product, Power of a Power, Product of Powers

14. Power of a Product, Power of a Power, Product of Powers

15. Answers may vary. Sample answer: Yes, the order makes a difference. Powers should be done before products.

Lesson 8.3

1. $\frac{-gh}{4}$ **2.** $\frac{-4x^2}{y^2}$ **3.** $\frac{p^3}{4q}$ **4.** $\frac{-5x^2y}{w^3}$

5. $\frac{-3q^5}{pr}$ **6.** $\frac{h}{2km}$ **7.** $\frac{-cd}{7}$

8. Answers may vary. Sample answer: The original and final expressions must have the same value for any values of the variables.

Lesson 8.4

1. $\frac{q^3}{p}$ **2.** $\frac{12}{x^4y^4}$ **3.** $\frac{3f}{e^3}$ **4.** $\frac{75d^2}{c^2}$ **5.** $\frac{-28c^3}{b^6d}$

6. $\frac{-4}{x^4y^2}$ **7.** $\frac{cd^3}{2a^4b}$ **8.** $\frac{-5x^5}{y^2}$ **9.** $\frac{-c}{7bd^3}$

10. Answers may vary. Sample answer: You cannot add exponents when the bases are not the same.

Lesson 8.5

1. 7.8×10^{21} **2.** 7.82×10^{-11}

3. 4.632×10^{13} **4.** 3.1×10^{-20}

5. 7.302×10^{15} **6.** 6.14×10^{-15}

7. $4,800,000,000$; 4.8×10^9

8. $32,900,000,000,000$; 3.29×10^{13}

9. 0.000084; 8.4×10^{-5}

10. 0.000000314; 3.14×10^{-7}

11. Answers may vary. Sample answer: First, change 2.4×10^9 to 24×10^8, or first change 1.68×10^8 to 0.168×10^9. Once the exponents are the same, you can add.

ANSWERS

Lesson 8.6

1. Yes; 2 is added to each *x*-value, and each *y*-value is multiplied by 4.

2. Yes; 1 is added to each *x*-value, and each *y*-value is multiplied by 0.4.

3. Yes; 3 is added to each *x*-value, and each *y*-value is multiplied by 0.1.

4. No; -1 is added to each *x*-value, but 13 is added to each *y*-value.

5. $y = 3 \cdot 2^x$

Lesson 8.7

1. 3   2. 5   3. 17   4. 257

5. Answers may vary. Sample answer: The numbers 3, 5, 17, and 257 are prime.

6. 4,294,967,297

7. No, 4,294,967,297 is not prime because dividing it by 641 results in the whole number 6,700,417.

8. $9 = 2 \cdot 2^2 + 1$

9. $25 = 3 \cdot 2^3 + 1$

10. $65 = 4 \cdot 2^4 + 1$

11. $5 \cdot 2^5 + 1 = 161$

12. $6 \cdot 2^6 + 1 = 385$

13. $7 \cdot 2^7 + 1 = 897$

14. Answers may vary. Sample answer: There are an infinite number of Cullen numbers because the number of integers is infinite.

Problem Solving/ Critical Thinking—Chapter 9

Lesson 9.1

1. like   2. unlike   3. like   4. unlike

5. unlike   6. unlike   7. like

8. unlike   9. like   10. like

11. $24x + 15x = 39x$

12. $18a^2 - 9a^2 = 9a^2$

13. $(6x^3 + 8) + 24x^3 = 30x^3 + 8$

14. $(4y^4 - 15) - (9y^4 + 5) = -5y^4 - 20$

15. $32a^6 + 30a^6 + 9a^6 = 71a^6$

16. Answers may vary. Sample answer: The general expressions $x^2 + x^2 + x^2$ and x^3 are not equal for numbers other than 3.

Lesson 9.2

1. difference of two squares

2. perfect square   3. neither

4. difference of two squares

5. perfect square

6. difference of two squares

7. neither   8. perfect square

9. difference of two squares   10. neither

11. $(a + b)(a + b) = a^2 + 2ab + b^2$; $9x^2 + 42x + 49$

12. $(a + b)(a - b) = a^2 - b^2$; $y^2 - 144$

13. $(a + b)(a - b) = a^2 - b^2$; $4x^2 - 25$

14. $(a + b)(a + b) = a^2 + 2ab + b^2$; $y^2 + 24xy + 144x^2$

15. $(m + n)(m + n)$; $m^2 + 2mn + n^2$

ANSWERS

Lesson 9.3

Check students' arrows.
1. $3x^2 + 10x + 3$

2. $2y^2 - 33y + 108$

3. $2x^2 + 11x - 40$

4. $y^2 + 8xy + 16x^2$

5. $8b^2 + 26bc + 15c^2$

6. $(2x + 7)(3x + 5)$; $6x^2 + 31x + 35$

Lesson 9.4

1. false

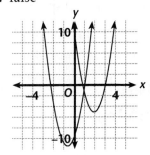

2. true

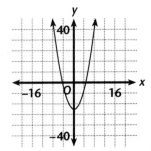

3. true

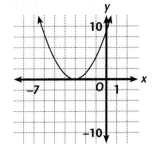

4. false

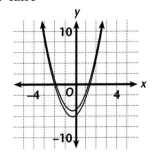

5. Answers may vary. Sample answer: The second graph may be out of the viewing range on the calculator.

Lesson 9.5

1. $4(ab^2 + 3a)$
   $4a(b^2 + 3)$

2. $5(5x^2 - 3x)$
   $5x(5x - 3)$

3. $17(c^4 + 3c^2)$
   $17c^2(c^2 + 3)$

4. $3(3w^3 + 4w^2 + 7w)$
   $3w(3w^2 + 4w + 7)$

5. $2(5y^6 + 4y^4 - 3y^2)$
   $2y^2(5y^4 + 4y^2 - 3)$

6. $5(mn^2 - 8m^2)$
   $5m(n^2 - 8m)$

7. $3(2xy^2z + 5xy)$
   $3x(2y^2z + 5y)$
   $3xy(2yz + 5)$

8. $a(19b^2c^3 - 35a^2b^2c)$
   $ab^2(19c^3 - 35a^2c)$
   $ab^2c(19c^2 - 35a^2)$

9. $4(c^5d^3 + 3c^2d)$
   $4c^2(c^3d^3 + 3d)$
   $4c^2d(c^3d^2 + 3)$

10. $11(2p^2r - 5pqr)$
    $11p(2pr - 5qr)$
    $11pr(2p - 5q)$

11. Answers may vary. Sample answer: Each expression is the GCF for the original polynomial.

ANSWERS

Lesson 9.6

1. $16y^2$ **2.** $81b^4$ **3.** $9g^2$ **4.** $25m^6$

5. 100 **6.** x^4 **7.** $2x, 4$

8. $5c^2, 12x$ **9.** $6a, c; 12ac = 2(6a)(c)$

10. $8x, 11y; 176xy = 2(8x)(11y)$

11. Answers may vary. Sample answer:
The middle term is not twice the product
of the roots. Stated mathematically:
$20xy \neq 2(5x)(4y)$

Lesson 9.7

1. 18 and 1, -18 and -1, 9 and 2, -9 and -2,
6 and 3, -6 and -3

2. 24 and 1, -24 and -1, 12 and 2, -12
and -2, 8 and 3, -8 and -3, 6 and 4,
-6 and -4

3. 12 and -1, -12 and 1, 6 and -2, -6 and 2,
4 and -3, -4 and 3

4. 30 and -1, -30 and 1, 15 and -2, -15
and 2, 10 and -3, -10 and 3, 6 and -5, -6
and 5

5. 15 and 1, -15 and -1, ⑤ and 3, -5
and -3

6. 12 and -1, -12 and 1, 6 and -2,
-6 and 2, 4 and -3, -4 and 3

7. 100 and -1, -100 and 1, 50 and -2, -50
and 2, ㉕ and -4, -25 and 4, 20 and -5,
-20 and 5, 10 and -10

8. 48 and 1, -48 and -1, 24 and 2, -24
and -2, 16 and 3, ⟨-16 and -3⟩ 12 and 4,
-12 and -4, 8 and 6, -8 and -6

9. 42 and -1, -42 and 1, 21 and -2, -21
and 2, 14 and -3, -14 and 3, 7 and -6,
⟨-7 and 6⟩

10. Answers may vary. Sample answer: Since
the middle term is positive, no two
negative numbers will have a sum equal to
its constant, a positive number.

Lesson 9.8

1. $x + 7 = 0, x - 9 = 0$

2. $n - 3 = 0, n - 8 = 0$

3. $5a - 2 = 0, 2a + 9 = 0$

4. $3y + 4 = 0, 4y + 5 = 0$

5. $x + 10 = 0, x + 2 = 0; x = -10$ and $x = -2$

6. $y - 5 = 0, y + 3 = 0; y = 5$ and $y = -3$

7. $r - 15 = 0, r - 2 = 0; r = 15$ and $r = 2$

8. $a + 12 = 0, a - 4 = 0; a = -12$ and $a = 4$

9. $d - 10 = 0, d + 9 = 0; d = 10$ and $d = -9$

10. Answers may vary. Sample answer: They
both cross the x-axis at the same points.

Problem Solving/ Critical Thinking— Chapter 10

Lesson 10.1

1. Answers may vary. Step d should show the
completed transformation shown below.

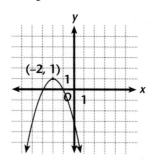

2. Answers may vary. Sample Answer: The
result should be the same regardless of the
order of the transformations.

Lesson 10.2

1. $x = \pm\frac{4}{5}$ **2.** $x = \pm\frac{8}{3}$ **3.** $x = \pm 5$

4. $x = \pm\frac{11}{4}$ **5.** $x = \pm\frac{1}{9}$ **6.** $x = \pm\frac{3}{2}$

ANSWERS

7. Answers may vary. Sample Answer: Divide both sides by the constant a. Then find the square root of both sides.

8. Answers may vary. Sample Answer: If $\frac{k}{a}$ is negative, then you cannot find its square root because no real number multiplied by itself gives a negative number. If $a = 0$, then $\frac{k}{a}$ is undefined.

Lesson 10.3

1. $y = (x - 4)^2 - 6$; vertex: $(4, -6)$; axis of symmetry: $x = 4$

2. $y = (x + 6)^2 - 9$; vertex: $(-6, -9)$; axis of symmetry: $x = -6$

3. $y = \left(x - \frac{5}{2}\right)^2 - \frac{1}{4}$; vertex: $\left(\frac{5}{2}, -\frac{1}{4}\right)$; axis of symmetry: $x = \frac{5}{2}$

4. $y = (x + 10)^2 - 1$; vertex: $(-10, -1)$; axis of symmetry: $x = -10$

5. $h = -\frac{b}{2}$ and $k = c - h^2$

Lesson 10.4

Student answers may be different.
1. factoring **2.** graphing

3. completing the square **4.** graphing

5. completing the square **6.** factoring

7. $x = 1$ or $x = -7$ **8.** $x = \frac{2}{3}$ or $x = -\frac{4}{3}$

9. no solution **10.** $x = 4$ or $x = 6$

11.
$$x^2 + bx + c = 0$$
$$\left(x + \frac{b}{2}\right)^2 + c - \frac{b^2}{4} = 0$$
$$\left(x + \frac{b}{2}\right)^2 = \frac{b^2 - 4c}{4}$$
$$x + \frac{b}{2} = \pm\sqrt{\frac{b^2 - 4c}{4}}$$
$$x = -\frac{b}{2} \pm \sqrt{\frac{b^2 - 4c}{4}}$$

Lesson 10.5

1. a. $x = 0.927$ and $x = -2.427$
b. $x = 0.771$ and $x = -2.271$
c. $x = 0.596$ and $x = -2.096$

2. a. $x = -0.5$ and $x = -2$
b. $x = -1$ and $x = -1.5$
c. no solution

3. $a = \dfrac{-6 + \sqrt{48}}{6}$ and $a = \dfrac{-6 - \sqrt{48}}{6}$;

$x = \dfrac{-(a + 1) \pm \sqrt{(a + 1)^2 - 4a(a + 2)}}{2a}$

If there is only one root, the expression under the square root symbol must equal 0. Solve $(a + 1)^2 - 4a(a + 2) = 0$.

$a = \dfrac{-6 \pm \sqrt{6^2 - 4(3)(-1)}}{2(3)}$, so the solutions

for a are $\dfrac{-6 + \sqrt{48}}{6}$ and $\dfrac{-6 - \sqrt{48}}{6}$.

Lesson 10.6

Test points and verifications may vary.
1.

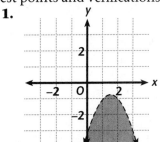

2.

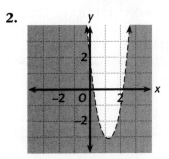

3.

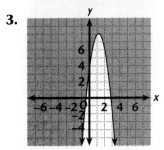

ANSWERS

4. Statement B is true. The solution of a quadratic inequality in x and y is one of two parts of the plane determined by the graph of a quadratic function. If (a, b) does not satisfy the inequality, then the solution region is the part of the plane that does not contain (a, b).

Problem Solving/ Critical Thinking— Chapter 11

Lesson 11.1

1. $k = 30$

2. $k = 32$

3. $k = 280$

4. $k = 240$

5. $y = \dfrac{36}{x}$, (1, 36), (2, 18), (3, 12), (4, 9), $\left(5, \dfrac{36}{5}\right)$

6. $y = \dfrac{240}{x}$, (1, 240), (2, 120), (3, 80), (4, 60), (5, 48)

7. $y = \dfrac{-90}{x}$, (1, −90), (2, −45), (3, −30), $\left(4, -\dfrac{45}{2}\right)$, (5, −18)

8. Answers may vary. Sample answer: The value of y approaches the constant of variation, k, as x approaches 1. The value of y approaches infinity as x approaches zero.

Lesson 11.2

1. $x = -5$; simple algebra

2. defined for all values; inspection

3. $x = 2$ or $x = -2$; factoring

4. $x = 4$ or $x = -7$; factoring

5. Anwers may vary. Sample answer: The domain will be the set of all real numbers.

Lesson 11.3

1. $\dfrac{b + 3}{2b^3 + 5}$

2. Cannot be simplified because the factors of the last term of the numerator must be 3 and −3, and neither of these factors is common to the last term of the denominator, −10.

3. $\dfrac{8x^2 + 3x + 7}{13x^3 - 6x}$

4. Cannot be simplified because the last terms share only the factors 1 and −1. Neither of these will give the correct middle term in the numerator.

5. Answers may vary. Sample answer: The variable is different in the numerator and denominator, so they can have no possible common factors.

Lesson 11.4

Answers may vary for fractions. Simplified expressions shown below.

1. $\dfrac{5x - 2}{6x}$

2. $\dfrac{4x + 13}{9x}$

3. $\dfrac{19x - 6}{20x}$

Lesson 11.5

1. $x = -1.5$ or $x = 5$

2. $x = -\dfrac{1}{2}$ or $x = 3$

3. $x = -5$

4. $x = -\dfrac{1}{7}$ or $x = 6$

5. Answers may vary. Sample answer: Because each of the denominators is a multiple of x, all of the expressions will be undefined when $x = 0$. This will eliminate the second possible solution from the solution set.

ANSWERS

Lesson 11.6

1. If a polygon is a triangle, then it has three sides.

2. If a polygon does not have three sides, then it is not a triangle.

3. If a polygon is not a triangle, then it does not have three sides.

4. converse, inverse, and contrapositive

5. If a polygon is quadrilateral, then it is a rectangle.

6. If a polygon is not a rectangle, then it is not a quadrilateral.

7. If a polygon is not a quadrilateral, then it is not a rectangle.

8. contrapositive

9. If two angles are equal in measure, then they are right angles.

10. If two angles are not right angles, then they are not equal in measure.

11. If two angles are not equal in measure, then they are not right angles.

12. None are true.

Problem Solving/ Critical Thinking— Chapter 12

Lesson 12.1

1. $7\sqrt{5} + 3\sqrt{2}$
   $+ 9\sqrt{5} + 2\sqrt{2}$

2. $3\sqrt{7} + 5\sqrt{3}$
   $- 2\sqrt{7} - 4\sqrt{3}$

3. $14\sqrt{3} - 5\sqrt{2}$
   $+ 8\sqrt{3} + 10\sqrt{2}$

4. $9\sqrt{2} - 3\sqrt{5}$
   $+ 3\sqrt{2}$
   $+ 6\sqrt{2}$

5. $17\sqrt{2} + 21\sqrt{3} + 33\sqrt{5}$
   $+ 3\sqrt{2} - 6\sqrt{3} - 3\sqrt{5}$

6. $16\sqrt{5} + 5\sqrt{2}$

7. $\sqrt{7} + \sqrt{3}$

8. $22\sqrt{3} + 5\sqrt{2}$

9. $18\sqrt{2} - 3\sqrt{5}$

10. $20\sqrt{2} + 15\sqrt{3} + 30\sqrt{5}$

11. Ordered pairs may vary. sample pairs: $(1, 0)$, $(0, 1)$, and $(2, 0)$; The solutions are $a = 0$ or $b = 0$.

Lesson 12.2

1. $x = 2$ 2. $x = 6$ 3. $x = 3$ and $x = 5$

4. $x = 8$ 5. $x = 4$ and $x = 2$

6. Explanations may vary. Sample answer: To solve $\sqrt{ax + b} = x$, solve $x^2 - (ax + b) = 0$, or $x^2 - ax - b = 0$.

Lesson 12.3

1.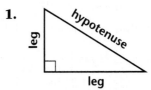

2.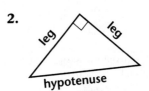

3.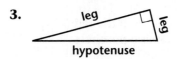

4. The unknown length of one leg equals the square root of the difference of the square of the hypotenuse and the square of the known length of the other leg.

ANSWERS

5. $a = 8$ 6. $a = 5$ 7. $a = 15$

8. Answers may vary. Sample answers:
   a. The formula was derived to solve for a leg of a right triangle not the hypotenuse.
   b. This triangle is not a right triangle, so the Pythagorean Theorem does not apply.

Lesson 12.4

1. point B; distance formula

2. point D; visually

3. point A; distance formula

4. Answers may vary. Sample answer: Put the point of the compass at the origin. Open the compass so that the other point is at A. Draw the circle. If point B is outside the circle, it is farther from the origin. If point B is on the circle, it is the same distance from the origin. If point B is inside the circle, it is closer to the origin.

Lesson 12.5

1. $(x + 1)^2 + (y - 2)^2 = 169$; no

2. $(x - 4)^2 + (y + 2)^2 = 100$; yes

3. $x^2 + (y - 3)^2 = 81$; no

4. $(x + 5)^2 + (y + 3)^2 = 289$; yes

5. $(x + 4)^2 + (y - 2)^2 = 64$; no

6. $(x - 1)^2 + (y - 1)^2 = 25$; yes

7. Answers may vary. Sample answer: If the value of the side of the equation containing x and y is less than the radius squared, then the point lies inside the circle. If it is more than the radius squared, the point lies outside the circle.

Lesson 12.6

1.
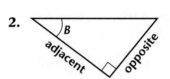

2.

3.

4. $\tan A = \frac{4}{3}$; $\tan B = \frac{3}{4}$

5. $\tan C = \frac{12}{5}$; $\tan D = \frac{5}{12}$

6. $\tan M = \frac{15}{8}$; $\tan N = \frac{8}{15}$

7. Answers may vary. Sample answer: The tangents of the two angles are reciprocals of each other.

Lesson 12.7

1.

2.
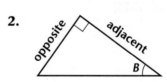

3.
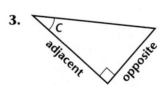

4. $\sin A = \frac{4}{5}$; $\cos A = \frac{3}{5}$

---

106   **Answers**

Algebra 1

ANSWERS

5. $\sin C = \frac{12}{13}$; $\cos C = \frac{5}{13}$

6. $\sin M = \frac{15}{17}$; $\cos M = \frac{8}{17}$

7. Answers may vary. sample answer:
   $\sin A = \cos B$

Lesson 12.8

1. $\begin{bmatrix} 61 & 103 & 119 \\ 127 & 114 & 145 \\ 151 & 85 & 126 \end{bmatrix}$

2. $\begin{bmatrix} 22 & 7 & 10 \\ 19 & 25 & 34 \\ 17 & 19 & 17 \end{bmatrix}$

3. $\begin{bmatrix} 87 & 90 & 15 \\ 169 & 249 & 94 \\ 98 & 139 & 56 \end{bmatrix}$

4. $\begin{bmatrix} 523 & 678 & 363 \\ 444 & 816 & 859 \\ 370 & 490 & 719 \end{bmatrix}$

5. $\begin{bmatrix} -4 & 8 & 14 \\ 0 & -12 & 13 \\ -3 & 8 & 0 \end{bmatrix}$

6. $\begin{bmatrix} -4 & 0 & -1 \\ 0 & 0 & 6 \\ -4 & -2 & -1 \end{bmatrix}$

7. The first and third matrices in the problem
   are the same as those in Exercise 6. So, their
   sum is already known. Subtract the middle
   matrix from this sum. The answer is
   $\begin{bmatrix} -6 & 1 & -4 \\ 2 & -1 & 9 \\ -6 & -1 & -1 \end{bmatrix}$.

Problem Solving/ Critical Thinking— Chapter 13

Lesson 13.1

1. $\frac{1}{2}$, or 50%

2. $\frac{4}{9}$, or 44.$\overline{4}$%

3. $\frac{5}{9}$, or 55.$\overline{5}$%

4. 8 centimeters × 8 centimeters

5. radius = 3 inches

6. horizontal leg = 12 inches

Lesson 13.2

1.

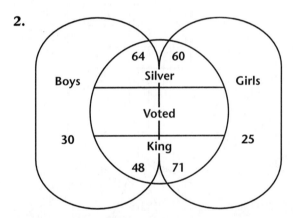

2.

3. 30 boys and 25 girls

Lesson 13.3

1. Fundamental Counting Principle

2. Fundamental Counting Principle

3. Addition Counting Principle

4. Fundamental Counting Principle

5. Answers may vary. Sample answer: As the
   number of consecutive days increase, the
   probability will get smaller.

ANSWERS

Lesson 13.4

1.
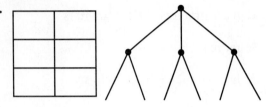

2.
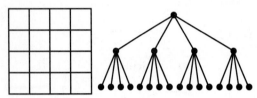

3.
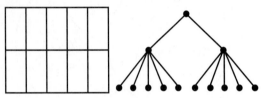

4.
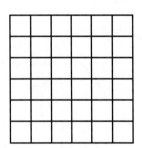

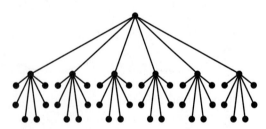

5. Answers may vary. Sample answer: Of the 36 possible outcomes, 6 of them add up to 7. This is more than any of the other totals, so the probability is higher.

Lesson 13.5

For Exercises 1–3, experimental probabilities will vary. Theoretical probabilities are given.

1. $\frac{1}{4}$

2. $\frac{1}{6}$

3. $\frac{1}{2}$

4. Answers may vary. Sample answer: As the number of trials increases, the experimental probability should get closer to the theoretical probability.

Problem Solving/ Critical Thinking— Chapter 14

Lesson 14.1

1. Not a function; element b in the domain is matched with two elements in the range.

2. Function; each element in the domain is paired with exactly one element in the range.

3. Not a function; element "gold" in the domain is paired with two elements in the range.

4. Answers may vary. Sample answer: At a single point in time, there is only one possible value for a measurement. You can never return to the same point in time to obtain a different measurement.

Lesson 14.2

1. $(-3, 5)$; 3 units left and 5 units up

2. $(2, -3)$; 2 units right and 3 units down

3. $(-6, -6)$; 6 units left and 6 units down

4. $(1.5, 4)$; 1.5 units right and 4 units up

5. $(-2.5, -4)$; 2.5 units left and 4 units down

6. $(2, -2.4)$; 2 units right and 2.4 units down

7. $(-3.8, 4)$; 3.8 units left and 4 units up

8. $(100, 100)$; 100 units right and 100 units up

9. $4(2^x) + 1 = 2^2(2^x) + 1 = 2^{x+2} + 1$; 2 units left and 1 unit up

ANSWERS

Lesson 14.3

Answers may vary. Sample answers:
1. *y*-values increase; the graph is narrower.

2. *y*-values decrease; the graph is wider and flatter.

3. *y*-values increase; the graph is narrower.

4. *y*-values increase; the graph is narrower.

5. *y*-values decrease; the graph is wider and flatter.

6. *y*-values increase; the graph is narrower.

7. *y*-values decrease; the graph is wider and flatter.

Lesson 14.4

Answers may vary. Sample answers:
1. *y*-values become their opposites and have smaller absolute values; the graph opens downward and is wider and flatter.

2. *y*-values become their opposites and have larger absolute values; the graph opens downward and is narrower.

3. *y*-values become their opposites and have larger absolute values; the graph opens downward and is narrower.

4. *y*-values become their opposites and have smaller absolute values; the graph opens downward and is wider and flatter.

5. *y*-values become their opposites and have smaller absolute values; the graph opens downward and is wider and flatter.

6. *y*-values become their opposites and have larger absolute values; the graph opens downward and is narrower.

Lesson 14.5

1. symmetric with respect to the *y*-axis

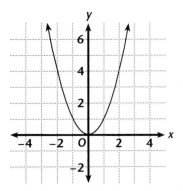

2. symmetric with respect to the *x*-axis

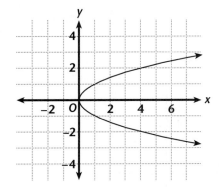

3. symmetric with respect to the *y*-axis, *x*-axis, origin, and the line $y = x$

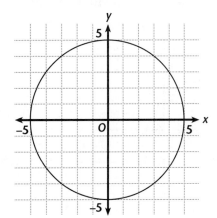

ANSWERS

4. symmetric with respect to the origin

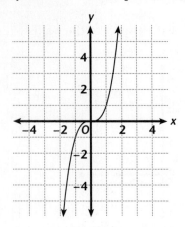

5. symmetric with respect to the y-axis

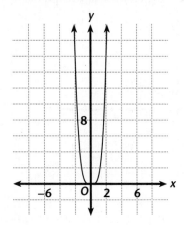

6. symmetric with respect to the x-axis

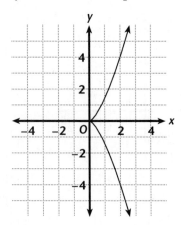